Multi-Criteria Decision Analysis via Ratio and Difference Judgement

Applied Optimization

Volume 29

The titles published in this series are listed at the end of this volume.

Multi-Criteria Decision Analysis via Ratio and Difference Judgement

by

Freerk A. Lootsma

Delft University of Technology,
Delft, The Netherlands

KLUWER ACADEMIC PUBLISHERS

DORDRECHT / BOSTON / LONDON

A C.I.P. Catalogue record for this book is available from the Library of Congress.

ISBN 0-7923-5669-1

Published by Kluwer Academic Publishers,
P.O. Box 17, 3300 AA Dordrecht, The Netherlands.

Sold and distributed in North, Central and South America
by Kluwer Academic Publishers,
101 Philip Drive, Norwell, MA 02061, U.S.A.

In all other countries, sold and distributed
by Kluwer Academic Publishers,
P.O. Box 322, 3300 AH Dordrecht, The Netherlands.

Printed on acid-free paper

Printed in the Netherlands.

CONTENTS

PREFACE

The point of departure in the present book is that the decision makers, involved in the evaluation of alternatives under conflicting criteria, express their preferential judgement by estimating ratios of subjective values or differences of the corresponding logarithms, the so-called grades. Three MCDA methods are studied in detail: the Simple Multi-Attribute Rating Technique SMART, as well as the Additive and the Multiplicative AHP, both pairwise-comparison methods which do not suffer from the well-known shortcomings of the original Analytic Hierarchy Process. Context-related preference modelling on the basis of psycho-physical research in visual perception and motor skills is extensively discussed in the introductory chapters. Thereafter many extensions of the ideas are presented via case studies in university administration, health care, environmental assessment, budget allocation, and energy planning at the national and the European level. The issues under consideration are: group decision making with inhomogeneous power distributions, the search for a compromise solution, resource allocation and fair distributions, scenario analysis in long-term planning, conflict analysis via the pairwise comparison of concessions, and multi-objective optimization. The final chapters are devoted to the fortunes of MCDA in the hands of its designers.

The research started in the late seventies, when I got involved in three different problems: the nomination procedures in a university, the evaluation of alternative energy-research proposals, and the evaluation of non-linear programming software. The decision makers who usually had a strong background in science and technology always questioned the scale sensitivity, the dependence on the units of measurement, the dependence on the methods employed in the experiments, and the rank preservation of the terminal scores. In the project teams we therefore thoroughly revised the original AHP which we employed since the late seventies. Using the Multiplicative AHP so obtained, and financially supported by the Ministry of Economic Affairs, we carried out several strategic-planning projects in close cooperation with the Energy Study Centre, North-Holland, during a period of fifteen years. In the late eighties we were also engaged in contract research commissioned by the Directorate-General XII (Science and Technology) of the European Community, Brussels, and by Schlumberger Research Laboratory, Paris. A generous grant of the Delft University Research Committee for the "Committee-Beek" project TWI 90-06

"Multi-Criteria Decision Analysis and Multi-Objective Optimization" enabled me to set up a large-scale project with a challenging goal: the development of robust MCDA methods for the allocation of scarce resources to the competing decision alternatives. In the nineties the project team has therefore been concerned with many issues such as the scaling of human judgement via geometric scales, comparative studies of current MCDA methods, extensions of MCDA methods to handle imprecise judgement, the power game in groups modelled via the assignment of power coefficients to the decision makers, and resource allocation via an appropriate ratio analysis. During the whole period, the International Institute of Applied Systems Analysis (IIASA), Laxenburg, Austria, provided excellent occasions for cooperation with research colleagues in East and West. I am also greatly indebted to the Department of Mechanical Engineering and Applied Mechanics, University of Michigan, Ann Arbor, for the freedom to work on fuzzy logic and multi-objective optimization during my sabbatical leave from Delft (1993 – 1994).

On a more personal level I would like to thank many people in chronological order. My first contact with energy problems was due to Harvey Greenberg (now at the University of Colorado, Denver) who invited me in 1977 to study the mathematical programming techniques in the PIES model developed at the US Department of Energy, Washington, DC. This also brought me into contact with the Energy Study Centre in Petten, North-Holland, and particularly with Frits van Oostvoorn. During the eighties we supervised many MSc projects, mostly in cooperation with Dick Mensch (Delft University of Technology) who is well-known for his careful style of working. Together with Manfred Grauer (IIASA, Laxenburg, Austria, now Universität-GH Siegen, Germany) we jointly supervised the PhD research by Matthijs Kok (Delft University of Technology, now HKV Lijn in Water, Lelystad, The Netherlands) who successfully implemented multi-objective optimization in the national energy model. In cooperation with Jan Meisner (Royal Shell/Dutch Research Laboratory, Amsterdam, now retired) we applied the first revision of the AHP for the selection of energy-research proposals, a study commissioned by the Energy Research Council, The Hague. A few years later, we were invited to carry out a similar study for the Directorate-General XII (Science and Technology) of the European Community, Brussels. Frans van Scheepen (DG XII, now retired) guided and supported us in an excellent manner, particularly when the project turned out to be rather delicate. Shouyang Wang (Chinese Academy of Sciences, Beijing), who spent two periods of one year in Delft, contributed significantly to our work on conflict analysis and negotiations. Throughout the eighties and the nineties, Warren Walker (RAND-Europe, Leiden, The Netherlands) sharpened my views on the decision phase of Operations Research projects by his experience in the analysis of large-scale infrastructural projects. Frans Vos (Delft University of Technology, now retired) developed the software and the questionnaires which we used for decision support in groups. With an iron discipline he also saved the data of our projects for numerical experiments in later years. His work was continued by Leo Rog (Delft University of Technology, now Nutsverzekeringen, The Hague) who contributed significantly to the "Committee-Beek" project by the development of the REMBRANDT system for MCDA. In later years the system was heavily tested by David Olson and the members of his project team (University of Texas, College Station, Texas) in a comparative study of several tools for MCDA. Bernard Roy (LAMSADE, Université de Paris-Dauphine) gave me several opportunities to work in his laboratory and to develop ideas which nevertheless diverge from his views on MCDA in general and on the position of the French school in MCDA. Joanne Linnerooth-Baier (IIASA, Laxenburg, Austria) highlighted the issues of fairness and equity, whereby she thoroughly affected the research direction in the "Committee-Beek" project. Josée Hulshof (Ministry of Health, Welfare, and Sport, The Hague, now Health Insurance Council, Amstelveen, The Netherlands), a strongly motivated and successful project manager, focussed my attention on the position of the chairperson in group decision making. She was very smoothly and effectively assisted by Pieter Bots (Delft University of Technology), facilitator in the Group Decision Room, and by Karien van Gennip (now McKinsey and Company, Amsterdam), who developed an unusual feeling for administrative problems when she was a student-member of the University Council in Delft. With Jonathan Barzilai (Dalhousie University, Halifax, Canada) I share the experience that it is hazardous to question the fundamentals of the established schools in MCDA. Panos Papalambros (MEAM, University of Michigan, Ann Arbor, USA), my host during my sabbatical leave in his department, gave me a generous opportunity to work on multi-objective optimization and fuzzy logic. Rob van den Honert (University of Cape Town, South-Africa) and R. Ramanathan (Indira Gandhi Institute of Development Research, Bombay, India) spent long periods in Delft as "Committee-Beek" Research Fellows in order to cooperate with me on group decision making and on issues of fairness and equity. Stan Lipovetsky (Tel Aviv University, Israel, now Custom Research, Minneapolis, Minnesota, USA) connected the geometric scales in MCDA with the experience of aesthetic pleasure via a generalization of the golden section. Finally, I am particularly grateful to the students Dionne Akkermans (now DSM Research, Geleen) and Marco van der Winden (now PGGM, Zeist) who tenaciously carried out several promising experiments in MCDA.

I dedicate this book to my wife Erica and to my children Joanne, Auke, and Roger, who kindly tolerated my absent-mindedness and who created the warm family atmosphere which drives away the loneliness of scientific research.

Delft, November 1998.

CHAPTER 1

INTRODUCTION

In the field of Multi-Criteria Decision Analysis (MCDA) we are concerned with the design of mathematical and computational tools to support the subjective evaluation of a finite number of decision alternatives under a finite number of performance criteria, by a single decision maker or by a group. Roughly thirty years old now, MCDA is a full-grown branch of Operations Research, if we take the number of publications in Operations Research Journals and the streams of contributions at the INFORMS and EURO conferences as yardsticks to judge its maturity.

1.1 ALTERNATIVES, CRITERIA, DECISION MAKERS

A decision is a choice out of a number of alternatives, and the choice is made in such a way that the preferred alternative is the "best" among the possible candidates. The decision process which preceeds the choice is not always very easy, however. Usually, there are several yardsticks to judge the alternatives and there is no alternative which outranks all the others under each of the performance criteria. Thus, the decision maker does not only have the task to judge the performance of the alternatives in question under each criterion, he/she also has to weigh the relative importance of the criteria in order to arrive at a global judgement. Moreover, in a group of decision makers each member faces the question of how to judge the quality of the other members and their relative power positions before an acceptable compromise solution emerges.

Many decisions take a long period of preparations, not only in a state bureaucracy or in an industrial organisation, but sometimes also in a small organisational unit like a family. As soon as a problem has been identified which is sufficiently mature for action, a decision maker is appointed or a decision-making body is established. The choice of the decision maker or the composition of the decision-making body usually emerges as the result of a series of negotiations where power is employed in a mixture of subtle pressure and brute force. The composition of the decision-making body reflects the strength or the influence of various parties in the organisation. In general, the members are also selected on the basis of their ability to judge at least some of the possible alternatives under at least some of the criteria. In many organisations there is a considerable amount of distributed decision making. In the mission of the decision-making committee the relevant criteria may have been prescribed and the relative importance of the criteria may have been formulated in vague verbal terms. So, the evaluation of the alternatives under the prespecified criteria is left to the experts in the committee, but the weighting of the criteria themselves is felt to be the prerogative or the resposibility of the authorities who established the committee. During the deliberations it may happen that new alternatives and/or new criteria emerge and that the composition of the decision-making body changes because new expertise is required. Nevertheless, there may be a clear endpoint of the decision process, in a particular session of the decision-making group where each member expresses his/her judgement. At this moment MCDA plays a significant role.

A typical example of such a decision process is the nomination procedure for a senior professorship, at least in our country. This is a lengthy procedure. It starts with a series of sometimes rather unstructured discussions about the desirable features of the vacancy, whereafter a Profile Committee is appointed with the task to draft a profile. This is in fact a list of vaguely described criteria to be satisfied by the future candidates. Note that the criteria are formulated and roughly weighted in the absence of applicants, henceforth in the absence of immediate context. Thereafter, the Profile Committee is modified and/or extended to become the Nomination Committee which has the task to advertise the vacant position, to interview the applicants, and to submit a short list with the names of the recommended candidates. The rank order on the list cannot be changed anymore, neither by the Faculty Council nor by any other Council or Board in the University which judges the proposal. Normally, the first-named candidate is appointed, but if he/she does not accept the vacant chair the Nomination Committee resumes the deliberations. The procedure may take two years, even if the profile was easily accepted MCDA can be very effective in the Profile Committee, when the profile is drafted, and also in the last meeting of the Nomination Committee, when the list of recommended candidates is made up, henceforth at eleven fifty five p.m. in the procedure.

The above process is fairly well structured: several universities have a scenario which clearly outlines the successive steps to be made by those who are involved in it. Many other decision processes in the public and the private sector are less transparent. The individuals who play a role in it do not really control the course of events so that they are sometimes referred to, not as decision makers, but as actors. The dynamic character of decision processes is somewhat overlooked in the literature on MCDA. It may happen in an early stage and at lower management levels that minor decisions are made which in a later stage seriously limit the number of alternative options for senior management. This prompted Howard Raiffa to state (ORSA/TIMS Conference, Boston, April 1994): "*Decisions are not always made, they just happen*". Thus, a major task for the chairperson of a decision-making body is to control when and where which step is made in the organisation. Moreover, the problem is not always to make a choice out of a given set of alternatives but to generate new alternatives via a series of negotiations in order to obtain an acceptable compromise solution for the committee.

1.2 OBJECTIVES OF MCDA

Methods for MCDA have been designed in order to designate a preferred alternative, to classify the alternatives in a small number of categories, and/or to rank the alternatives in a subjective order of preference. As we will see in the present volume, they may sometimes also be used to allocate scarce resources to the alternatives on the basis of the results of the analysis. Scanning the literature (Keeney and Raiffa, 1976; Saaty, 1980; Roy, 1985; Von Winterfeldt and Edwards, 1986; French, 1988; Vincke, 1989; Stewart, 1992; Roy and Bouyssou, 1993) we find that MCDA is usually supposed to have some or all of the following objectives:

1. Improvement of the satisfaction with the decision process. MCDA urges the decision makers to frame the decision problem and to formulate the context explicitly. Next, MCDA structures the problem because the decision makers are requested to list the alternatives and the criteria, and to record the performance of the alternatives under each of the relevant criteria, either in their original physical or monetary units or in verbal terms. This step alone, the screening phase which is full of decision traps (Russo and Schoemaker, 1989), is frequently felt to be the most important contribution of MCDA because the systematic inventarization presents a global overview of the possible choices and their consequences. Moreover, MCDA aids the decision makers in the formulation of the criteria because it shows the priorities and the values which may be deeply hidden in the back of their mind. In fact, it makes the criteria operational. MCDA also supports the decision makers in the evaluation of the alternatives because it shows the subjective values of the performance of the alternatives within the context of the decision problem. Finally, MCDA eliminates or reveals the hidden objectives of certain members in a group (the hidden agenda), and it reduces the effects of certain discussion techniques. It reduces the dominant role of the members with strong verbal skills, for instance, so that the silent majority has a proper chance to weigh the pros and cons of the alternatives and to insert their judgement in the decision process. In short, MCDA enhances the communication in the group.

2. Improvement of the quality of the decision itself MCDA enables the decision makers to break down a decision problem into manageable portions and to express a detailed

judgement. The decision makers are not easily swept away by the performance of some alternatives under one or two criteria only, but they keep an eye on the performance of all alternatives under all criteria simultaneously. Moreover, MCDA may suggest a compromise solution in a group of decision makers, possibly after several rounds of discussions, with due regard to the relative power positions of the group members. It is our experience that MCDA may come up with an attractive proposal in an early stage of the decision process. The decision makers, however, need several rounds of discussion before they accept the proposed alternative.

3. Increased productivity of the decision makers: more decisions per unit of time, both in public administration and in industrial management. This objective ranks high on the agenda of decision makers who are repeatedly involved in procedures to evaluate the performance of personnel, development projects, and research proposals. The criteria and the alternatives are rather similar, from one session to the next, so that MCDA could be used to save time and energy.

MCDA also has some drawbacks. It introduces a formalised style of working, possibly in cooperation with an analyst and a facilitator using a computer or even a network of computers. This constitutes an extra burden for a decision maker, and particularly for (the chairperson of) a group of decision makers. And what is the trade-off between the increased effort on the one hand and the improved satisfaction with the decision process, the improved quality of the decision, and/or the increased productivity of the decision makers on the other?

1.3 RESEARCH IN MCDA

The research in the field of MCDA follows a number of distinct approaches in order to support the decision makers in their attempts to identify a preferred alternative, to classify the alternatives, and/or to rank-order them.

1. The descriptive approach merely tells how decision makers actually behave when they are confronted with the choice between several alternatives under conflicting viewpoints and how MCDA contributes to the decision process. Many studies are concerned with individual and collective decision making, with an analysis of the rationality of decision makers in various cultural contexts, and with the identification of hidden objectives. Many specialists in this research area hesitate to believe that MCDA is compatible with the style of decision making of human beings. The rigidity and the formality of MCDA are mostly in conflict with the heuristic tactics applied by human beings during the decision process.

2. The normative approach tells how decision makers should behave and how MCDA should work, via logical rules which are based upon certain fundamental axioms such as the transitivity of preferences. The typical products of the approach are Multi-Attribute

Value Theory (MAVT) and Multi-Attribute Utility Theory (MAUT), to be used for decision problems with certain and with uncertain outcomes respectively. The key element of MAVT is the concept of a value function which represents the degree to which the alternatives satisfy the objective of the criterion under consideration. In MAUT the user is requested to construct a utility function by choosing the monetary aequivalent for a lottery or the aequivalent lottery for a given amount of money. Despite the firmness of their axiomatic foundations these methods are not easy to use. The verbal probability estimates expressed by human beings, for instance, seem to be notoriously poor. Furthermore, the rational axioms of preference judgement are not always obeyed by human beings. Cyclic judgement, for instance, can easily occur under qualitative criteria, when human beings carry out a dynamic search for a proper perspective on the alternatives.

3. The prescriptive or clinical approach tells how decision makers could improve the decision process and the decisions themselves, and how MCDA could support such a process Key elements in the approach are the modelling of human judgement, the identification of preference intensities, the aggregation procedures, and the design of decision-support systems. The clinical approach can typically be regarded as a branch of Operations Research because it is also concerned with ad hoc models and algorithmic operations in order to support actual decision making. On the one hand, it may be successful when it concentrates on the needs of the users and on the potential benefits of mathematical analysis and information processing. On the other hand, it is always in danger when it ignores the volatility and the hidden agenda of the decision makers.

4. The constructive approach questions the existence of a coherent, well-ordered system of preferences and values in the decision maker's mind, as well as the idea that MCDA should correctly fathom such a pre-existing system. These are the more or less tacit assumptions underlying the normative and the prescriptive approach. The constructivists, however, advocate that the decision maker and an analyst should jointly construct a model of the system, at least as far as it is relevant in the actual situation. In other words, the decision maker's preferences and values, initially unstable and volatile or even non-existent, will be shaped by MCDA. This implies that the results of the analysis may be highly dependent on the analyst and on the method which he/she employs.

This short summary tends to be an oversimplification of the basic philosophies behind the respective approaches. For more details of the descriptive, the normative, and the prescriptive approach we refer the reader to Von Winterfeldt and Edwards (1986), Bell et al. (1988), and French (1988). The constructive approach is followed by the French school in MCDA. It has extensively been described by Roy (1985) and Roy and Bouyssou (1993) They usually refer to the normative approach as the American school in MCDA Recently, Roy and Vanderpooten (1996) declared the French school to be the European school in MCDA, a statement which we refuted in the comment following their article. It is unclear whether the methods which have been designed in the French and the American school respectively satisfy the typical needs of the French and the American decision makers. Research in MCDA is usually concerned with various methods within the framework of the respective approaches, not with the differences between styles of management and decision making in various parts of the world.

Nevertheless, constructivism raises several questions which cannot be ignored. What does MCDA produce, for instance? Does it bring up, faster and with more satisfaction, the preferred alternative which would emerge anyway, whether the decision maker is supported by MCDA or not? Or does it generate something which is totally new? And how reasonable are the more or less tacit assumptions underlying the applications of MCDA in groups? There are methods which calculate averages of weights and scores in order to approximate the common preferences and values in the group, in attempts to guide the group towards consensus. These common preferences and values may indeed be present when the group members belong to the same organisation. The information produced by applications of MCDA may alternatively reveal the diverging clusters of preferences and values in the group. The pattern of clusters may be used by the group members to form a majority coalition. And how does MCDA incorporate the ongoing power game in groups? Does it enable the members to arrive at a consensus which is compatible with the balance of power? These questions are rarely discussed in the literature on MCDA. In the present volume we will sketch several variants of MCDA methods which support consensus seeking or the formation of coalitions.

1.4 TAXONOMY OF DECISIONS AND DECISION MAKERS

Numerous multi-criteria decisions are daily made, both in public and in private life: strategic decisions (in a company the choice of products and markets, for instance, and in private life the choice of a partner and a career), tactical decisions (the choice of a location for production and sales, the choice of a university or a job), and operational decisions (daily or weekly scheduling of activities). Numerous decision makers are also involved in them: charismatic leaders, cool administrators, and manipulating gamesmen, for instance, who all (inconsistently?) adopt widely different tactics in their style of decision making. Sometimes they defer the decisions to higher authorities, sometimes they delay a decision until there is only one alternative left or until the problem evaporates, and sometimes they deliberate the pros and cons of the alternative options before they arrive at a conclusion. Some decision makers are experienced (the physicians who are specialized in the treatments they prefer), others are totally unexperienced (the patients who have to choose a particular treatment for themselves). And although decision makers are usually not illiterate, some of them seem to be innumerate in the sense that they are insensitive to what numerical values mean within a particular context. Many decisions are made in groups by members who may have widely varying power positions. On certain occasions a brute power game seems to be acceptable, on other occasions the members of the group moderate their aspirations, out of self-interest or motivated by considerations of fairness and equity.

Cultural differences cannot be ignored in decision making. Culture is the collective programming of the mind which distinguishes the members of one group from another (Hofstede, 1984). It is the way in which a group of people solves the universal dilemmas of human existence: the relationships with other human beings, with time, and with the natural environment (Trompenaars, 1993). Particular dimensions for a cross-cultural comparison of the relationships with other human beings are: the power distance indicated by the fear to express disagreement with superiors, uncertainty avoidance indicated by rule orientation and employment stability, individual self-realisation versus the pursuit of collective interests, status achieved on the basis of performance and skills versus status ascribed on the basis of family background and seniority, and neutral behaviour versus a vivid expression of feelings (Hofstede, 1984). In some cultures the group may leave the decision to the highest ranking member, in others the group may work towards a maximal consensus or towards a majority standpoint which overrules the viewpoints of the minorities. The border of the Roman Empire still is a clear-cut border between North-West Europe (rule-based and task-oriented behaviour, status achieved) and Latin Europe (person-oriented and affective behaviour, status ascribed). There is a second, but less deeper divide in the Western world: the gap between the Germanic culture (rather hierarchical, organisations symbolized as well-oiled machines) and the Anglo-Saxon culture (more egalitarian, organisations symbolized as markets). The series of case studies in Trompenaars (1993) illustrates how deeply cultural differences may affect decision processes.

With such an overwhelming variety of decisions and decision makers it is difficult to believe in a widespread applicability of the standard problem formulation of MCDA (the choice out of a finite set of alternatives under a finite set of conflicting criteria). And indeed, many decision processes go through a number of cycles at various levels in the organisation. Decision makers are replaced, new alternatives are generated via a series of negotiations, old ones are dropped, and new criteria emerge, until the group arrives at a consensus. We discussed this in Section 1.1.

1.5 PREFERENCE MODELLING

In the present volume we shall be concerned with a restricted number of methods for MCDA. Because human beings usually express their preferences in terms which reveal gradations of intensity (indifference, weak, definite, strong, or very strong preference), we limit ourselves to cardinal methods on the assumption that the preference information which is obtained in the respective elicitation procedures constitutes ratio or difference information. We shall occasionally turn to ordinal methods where the decision makers merely rank-order their preferences. Our starting point is basically an observation made by Torgerson (1961): in a pairwise comparison of two stimuli on a one-dimensional scale, human beings perceive only one quantitative relation between the subjective stimulus values, but they estimate ratios (differences) when a category scale with geometric (arithmetic) progression is presented to them. Thus, they interpret the relation as it is required in the experiment. Which of the two interpretations, ratios or differences, is correct cannot empirically be decided because they are alternative ways of saying the same things. Any model which relies solely on ratios can be parallelled with a model which relies solely on differences. Thus, there is a logarithmic relationship between the ratio and the difference information in the MCDA methods to be considered here: ratios of subjective values may be recorded as differences of grades. In psycho-physics this is known as logarithmic coding (Bruce and Green, 1990). It can typically be observed in the field of acoustics where ratios of sound intensities are usually recorded as differences on a logarithmic scale, the well-known decibel scale. We henceforth concentrate on two methods.

1. The Simple Multi-Attribute Rating Technique (SMART). The performance of the alternatives under the respective criteria, evaluated via a direct-rating procedure, is expressed in grades on a numerical scale.

2. The Analytic Hierarchy Process (AHP). The alternatives are considered in pairs. Their relative performance can aequivalently be expressed as a ratio of subjective values (Multiplicative AHP) or as a difference of grades (Additive AHP, or SMART with pairwise comparisons).

In order to model the preferences of the decision maker we assume that decisions are invariably made within a particular context. Under a quantitative criterion, where the performance of the alternatives can be expressed in physical or monetary units, we model the context as the range of acceptable performance data on the corresponding dimension. The lower and the upper bound are to be estimated by the decision maker. Next, starting from a particular viewpoint we partition each of these ranges into a number of subintervals which are subjectively equal. On the basis of psycho-physical arguments the gridpoints demarcating the subintervals constitute a geometric sequence. The behaviour of the gridpoints is in fact surprisingly uniform. The progression of historical periods and planning horizons, the categorization of nations on the basis of the population size, and the categorization of light and sound intensities, they all have a progression factor which is roughly equal to 2. This enables us to categorize the ranges and to quantify verbal expressions like indifference as well as weak, definite, and strong preference, the so-called gradations of relative preference. Thus, the leading idea of the approach is that the decision maker, standing at the so-called desired target, views the range of acceptable performance data in perspective. Under the qualitative criteria, where the performance of the alternatives can only be expressed in verbal terms, we use these results in an analogous manner.

In essence, the measurement of preference intensities is closely related to the subjective measurement of space and time, light and sound. When human beings express their preferences, they measure the performance of the alternatives, that is, they estimate what the alternatives will afford them to do and/or to enjoy. They have powerful tools to do so because subjective measurement is an integral part of their daily life. They continually perceive the surrounding environment and themselves in it. They continually measure the past, present, and future position of the objects around them with respect to their body. Human beings are dynamic spectators, however. They perceive in order to move, but they also move in order to perceive (Gibson, 1979; Kelso, 1982). This is shown by the visual effects of the so-called retinal stabilization, but the experiment can also be carried out in a simple manner, without sophisticated equipment. One just has to keep the eyes in a fixed position and one only has to concentrate on a particular object in order to see that the image gradually disappears. It is also a common experience that, in order to perceive an object, one has to approach it, one has to walk around it, and one has to observe it under different angles. The impulse is irresistible. Thus, one moves in order to perceive the invariant features in the changing perspective, and one learns to see things in perspective by the development of the so-called motor skills.

Space is more or less tangible and/or visible, but time and preference are volatile. Living creatures have a surprising ability, however, to control a time-dependent series of rhythmic actions like walking, running, and tapping, which are controlled by a timekeeper. Many living creatures also have a biological or physiological clock to measure the time which elapsed since a particular moment. So, if time can subjectively be measured, the gradations of preference may be measurable as well. We therefore model the gradations of preference on the basis of the working hypothesis that the gradations are shaped by the decision maker's perspective on the range of acceptable performance data. The ability to see the surrounding environment in perspective is associated with his/her motor skills.

Quantitative criteria are usually one-dimensional: the performance of the alternatives can be expressed in physical or monetary units on a one-dimensional scale. Qualitative criteria, however, are often low-dimensional: the performance of the alternatives can only be expressed in several more or less verbal terms. The dynamic perception of the alternatives and the search for a proper perspective may easily lead to incomparability, intransitivity, and cylic judgement of the alternatives. These phenomena are not necessarily irrational for dynamic spectators in a low-dimensional space. Many decision makers may have the experience that they are trapped in a cycle when they are confronted with important moral or ethical aspects in a decision problem.

1.6 SCOPE OF THE PRESENT VOLUME

This volume is organized as follows. The Chapters 2 – 5 present the basic ideas of SMART and the AHP. In the Chapters 6 – 10 we consider extensions which are also based upon the assumption that the decision makers supply ratio and/or difference information. Miscellaneous subjects are discussed in the Chapters 11 and 12.

Basic ideas. In Chapter 2 we turn to SMART. We describe the assignment of impact grades to the alternatives under the respective criteria, the assignment of weights to the criteria, and the generation of the final grades of the alternatives. The last-named operation is carried out via an arithmetic-mean aggregation rule. Throughout this chapter we show that the decision maker can work with a mix of scales: a numerical, a verbal, and a symbolic one, or just with ordinal information. Examples to illustrate the model formulation are given by the selection of a car, the evaluation of infrastructural projects, and the quality assessment of academic research. In Chapter 3 we consider the Multiplicative and the Additive AHP jointly. We show that the two methods have the same algorithmic core: the impact scores of the alternatives under a particular criterion are calculated via the solution of the same least squares problem, regardless of whether the decision maker supplied ratio or difference information. The procedure can easily be generalized in order to deal with group decision making. In the aggregation procedure we obtain the terminal scores of the alternatives via a sequence of arithmetic-mean calculations (Additive AHP) or geometric-mean calculations (Multiplicative AHP). Thus, the terminal scores do not depend on the order of the operations. The AHP can also easily deal with incomplete and intransitive pairwise comparisons. We analyze whether cyclic judgement is irrational and whether an extended hierarchy of criteria and subcriteria would "solve" the paradox of cylic judgement (it does not). We illustrate the approach via nomination procedures in a university and staff appraisal in industrial research. Although geometric scales to quantify human comparative judgement seem to be plausible, the establishment of the progression factor via the perception of time, space, light, and sound remains questionable. In Chapter 4 we therefore thoroughly analyze the scale sensitivity of the terminal scores under reasonable variations of the progression factor. Numerical experiments with small and large test problems, such as the evaluation of diseases and energy technologies, show that the sensitivity is surprisingly low. Moreover, there is a simple proof to show that the rank order of the terminal scores is preserved under variations of the progression factor over a positive range of values. Chapter 5 is concerned with preference modelling and with the decision maker's dynamic perspective on the alternatives. In essence, this chapter is an intermezzo which sketches the physiological and philosophical background of MCDA, see also Section 1.5.

Extensions. Chapter 6 is devoted to the potential of SMART and the AHP in group decision making. We first consider the facilities for weighted judgement whereby we model inhomogenuous distributions of power and influence, and we illustrate the results with the seat allocation in the European Parliament. Next, we analyze various procedural strategies to achieve consensus. The rank order of the terminal scores appears to depend on the order of the procedural steps in the group meeting. The chairperson may ask the group members to evaluate the alternatives and the criteria independently, and thereafter he/she may combine the terminal scores into a group standpoint. He/she may also follow a course whereby the group is invited to achieve consensus at intermediate stages. In that case the group works with compromise criterion weights and with compromise impact grades and scores. The choice of the procedural strategy affects the values and the rank order of the terminal scores. This is presumably well-known among experienced chairpersons. In Chapter 7 we allocate scarce resources to competing alternatives such as energy research programs on the basis of the terminal scores. Using the leading principles of fairness and equity (proportionality, progressivity, priority, and parity) we calculate the contributions to be paid by the member states to a union or a federation on the basis of several distribution criteria such as the size of the population and the Gross Domestic Product. The European Union provides a typical field of applications for the approach. Chapter 8 deals with decision problems where the performance of the alternatives strongly depends on future circumstances. In a booming economy, for instance, certain expensive energy technologies may be welcome, but the capital investments may be a clog to the legs in a long period of recession. In general, we are confronted here with the choice between various strategies within the framework of several (more or less plausible) scenarios of future events which are beyond the decision maker's control. We discuss the analysis of several long-term strategies in a project to support the choice between various energy carriers for electricity production: gas, coal, oil, and uranium. There were many delicate political issues to be considered. In order to avoid the pitfalls the group accepted a solution which had been generated via a series of negotiations. Chapter 9 shows that ratio information is crucial in the attempts to reduce or to solve a conflict via pairwise comparison of the possible concessions that could be made by the two parties. In Chapter 10 we turn to the class of multi-objective optimization problems where we have a continuum of alternatives defined by a number of constraints There are in fact two subfields here: the identification of the non-dominated solutions and the selection of a non-dominated solution for actual implementation. The identification problem can be solved in the splendid isolation of mathematical research, but subjective human judgement is an integral part of the selection process MCDA handles ratio information in order to support the activities in the second subfield.

Miscellaneous subjects. Chapter 11 is devoted to the author's experiences in actual applications of MCDA, not only to the successes, however, but also to the failures which were highly instructive. A surprising phenomenon was the inability or the unwillingness of the specialists in MCDA to use their own methods in order to solve their own decision problems. This could be observed, for instance, in the attempts to choose a combination of working languages in a European Working Group for Multi-Criteria Decision Aid. The course of the events may have been typical for the European unification, however, so that the disappointing results should be taken to indicate a delicate issue with deeper roots. In a speculative fashion, Chapter 12 is finally concerned with the future of MCDA and the tools for decision support.

1.7 THE ASSIGNMENT OF GRADES

An important and controversial step in MCDA is the conversion of more or less objective and quantitative information (the performance of the alternatives in physical or monetary units) into numerical data (the impact grades and scores) expressing subjective and qualitative human judgement. The conversion into judgemental data is a familiar step, however, since we have all been subject to the assignment of grades in schools and universities. Many specialists in MCDA were not only subject to it: in the role of faculty staff members they also had to assign grades themselves. Let us therefore recall here the underlying assumptions and ideas before we start with the presentation of MCDA methods. What do we have to consider before we proceed to the assignment of grades? What do we plan to achieve and how valuable will the grades be?

1. Framing is the first step. The possible performance of the students has to be considered within a particular context. The exercises submitted to them must be compatible with what they are supposed to know, to understand, and/or to do within the context of an educational program. At any stage there is a maximum level, perhaps attainable for some of the students only, and there is a minimum level, hopefully attainable for almost all students in the group to be examined. As soon as these levels have been determined, that is, as soon as the scale for the teacher's judgement has been anchored, the assignment of grades is a matter of interpolation. The determination of the levels is not an easy task, however. It should be carried out in consultation with colleagues in adjacent fields and with a view on the educational program. In order to do it successfully one needs several years of experience and cooperation.

2. Grades are used for any subject. In mathematics, for instance, one may just count the number of errors in order to measure the performance of the students and to assign the grades. In the field of classical or modern languages one may directly assign grades to essays, for instance. Although there are quantitative and qualitative subjects (criteria), there is a uniform scale to record the performance of the students and to support pass-or-fail decisions at the end of an educational period.

3. Usually, grades are numerical values on a scale with equidistant echelons, a so-called arithmetic scale like 1, 2, ..., 10, with 10 for excellent performance, 6 for fair or just sufficient performance, and 4 for unsatisfactory performance that can be compensated elsewhere in pass-or-fail decisions. So, there is at least a seven-point scale with equal distances between the echelons. The grades 1, 2, and 3 usually designate a (very) poor performance that cannot be compensated. Each echelon on the scale represents a performance category. Equal steps along the scale do not necessarily represent equal variations in performance. On the contrary, the performance varies superlinearly or exponentially with moves along the scale.

4. It may happen that the subjects or courses have unequal weights. In the group of modern languages, for instance, English is usually more important than French and German. In an educational program leading to a degree in science and/or engineering, mathematics is one of the most important subjects. It will be obvious that the weights do not depend on the performance of the students but on external factors such as the societal importance of the subjects. The weights are normally used to calculate weighted arithmetic means of judgemental data. Thus, they operate on the grades assigned to the students, not on the original data such as the number of errors in the

exercises. By this mode of operation one does not have to judge the trade-off between the number of errors in mathematics and in an essay, for instance. One only has to consider the quality of the students in various subjects or courses. The concept of the relative importance of the subjects can now easily be illustrated. If there are two applicants for a fellowship, for instance, one of them with the grades 9 and 7 for mathematics and the national language respectively, and the other with the grades 7 and 9 for the same subjects, then the choice depends on the nature of the fellowship and on the relative importance of the subjects within it.

5. For the teacher, who is normally one of the referees or one of the decision makers in collective pass-or-fail decisions, the assignment of grades is not a matter of purely individual taste. The teacher is not completely free in the assignment because he/she operates in a social and professional context. He/she may be requested by the dean of the school or by his/her colleagues to account for the grades assigned to his/her students. Thus, grades are more or less public, and the pass-or-fail decision must be transparent.

We suggest the readers to keep these considerations in mind when they are concerned with the chapters to follow. The assignment of grades has been current practice during many centuries. Although we do not assert that it is a perfect system, it has extensively been used for the evaluation of alternatives (students) under a variety of criteria (subjects or courses), and the decisions to be made were important.

REFERENCES TO CHAPTER 1

1. Bell, D., Raiffa, H., and Tversky, A. (eds.), *"Decision Making: Descriptive, Normative, and Prescriptive Interactions"*. Cambridge University Press, Cambridge, Mass., 1988.

2. Bruce, V., and Green, P.R., *"Visual Perception, Physiology, Psychology, and Ecology"*. Lawrence Erlbaum, Hove, UK, 1990.

3. French, S., *"Decision Theory, an Introduction to the Mathematics of Rationality"*. Ellis Horwood, Chichester, UK, 1988.

4. Gibson, J.J., *"The Ecological Approach to Visual Perception"*. Houghton-Mifflin, Boston, 1979.

5. Hofstede, G.H., *"Culture's Consequences, International Differences in Work-related Values"*. Sage, Beverly Hills, 1984.

6. Keeney, R., and Raiffa, H., *"Decisions with Multiple Objectives: Preferences and Value Trade-Offs"*. Wiley, New York, 1976.

7. Kelso, J.A.S., *"Human Motor Behavior, an Introduction"*. Lawrence Erlbaum, Hillsdale, New Jersey, 1982.

8. Lootsma, F.A., *"Fuzzy Logic for Planning and Decision Making"*. Kluwer Academic Publishers, Boston, 1997.

9. Roy, B., *"Méthodologie Multicritère d'Aide à la Décision"*. Economica, Collection Gestion, Paris, 1985. *"Multicriteria Methodology for Decision Aiding"*. Kluwer Academic Publishers, Dordrecht, The Netherlands, 1996.

10. Roy, B., et Bouyssou, D., *"Aide Multicritère à la Décision: Méthodes et Cas"*. Economica, Collection Gestion, Paris, 1993.

11. Roy, B., and Vanderpooten, D., "The European School of MCDA: Emergence, Basic Features, and Current Works". *Journal of Multi-Criteria Decision Analysis* 5, 22 – 37, 1996. In the same volume there is a comment by F.A. Lootsma (37 – 38) and a response by the authors (165 – 166).

12. Russo, J.E., and Schoemaker, P.J.H., *"Decision Traps"*. Simon & Schuster, New York, 1989.

13. Saaty, T.L., *"The Analytic Hierarchy Process, Planning, Priority Setting, and Resource Allocation"*. McGraw-Hill, New York, 1980.

14. Stewart, T.J., "A Critical Survey on the Status of Multiple Criteria Decision Making Theory and Practice". *Omega* 20, 569 – 586, 1992.

15. Torgerson, W.S., "Distances and Ratios in Psycho-Physical Scaling". *Acta Psychologica* XIX, 201 – 205, 1961.

16. Trompenaars, F., *"Riding the Waves of Culture, Understanding Cultural Diversity in Business"*. Brealey, London, 1993.

17. Vincke, Ph., *"L'Aide Multicritère à la Décision"*. Editions de l'Université Libre de Bruxelles, Belgique, 1989.

18. Winterfeldt, D. von, and Edwards, W., *"Decision Analysis and Behavioral Research"*. Cambridge University Press, Cambridge, UK, 1986.

CHAPTER 2

SMART, DIRECT RATING

The Simple Multi-Attribute Rating Technique (SMART, Von Winterfeldt and Edwards, 1986) is a method for Multi-Criteria Decision Analysis (MCDA) whereby we evaluate a finite number of decision alternatives under a finite number of performance criteria. The purpose of the analysis is to rank the alternatives in a subjective order of preference and, if possible, to rate the overall performance of the alternatives via the proper assignment of numerical grades. In this chapter we present SMART in its deterministic form, regardless of the vagueness of human preferential judgement. As a vehicle for discussion we use the example which is frequently employed to illustrate the applications of MCDA: the evaluation and the selection of cars.

2.1 THE SCREENING PHASE

MCDA starts with the so-called screening phase which proceeds via several inventarizations. What is the objective of the decision process? Who is the decision maker or what is the composition of the decision-making body? What are the performance criteria to be used in order to judge the alternatives? Which alternatives are in principle acceptable or not totally unfeasible?

These questions are not always answered in a particular order. On the contrary, throughout the decision process new alternatives may appear, new criteria may emerge, old ones may be dropped, and the decision-making group may change. Nevertheless, these steps in the process are inevitable. Many decision problems are not clear-cut, and the decision makers have to find their way in the jungle of conflicting objectives (see also Russo and Schoemaker, 1989).

When a family selects a car, these features of the decision process also emerge. First, the members have to identify the problem. Do we need to replace the old car? Do we only consider cars, or do we also take a public-transportation card as a feasible alternative to solve the transportation problems of the family? This question must be answered before any further progress can be made. The choice between public transportation and a car is not merely a choice between vehicles but a choice between life-styles.

Next, how to judge the cars? On the basis of generally accepted criteria that other people normally use as well? Can we also use our experiences in the past to introduce new criteria? Are there any particular cars on the market which lead to new criteria? Do we only want to compare the cars themselves or do we also consider the supporting dealer networks, both on the home market and abroad? And who are the decision makers? The parents only?

The result of the screening phase is the so-called performance tableau which exhibits the performance of the alternatives. Under the so-called quantitative or measurable criteria the performance is recorded in the original physical or monetary units. Under the qualitative criteria it can only be expressed in verbal terms. Table 2.1 shows such a possible tableau for the car selection example The tacit assumption is that the alternatives are in principle acceptable for the decision makers and that a weak performance under some criteria is compensatable by an excellent performance under some of the remaining ones. In other words, the decision makers are in principle prepared to trade-off possible deficiencies of the alternatives under some criteria against possible benefits elsewhere in the performance tableau. The alternatives which do not appear in the tableau have been dropped from consideration because their performance under at least one of the criteria was beyond certain limits. They were too expensive, too small, or too slow, for instance.

Table 2.1 *Performance tableau of four alternative cars under ten criteria.*

Criterion	Unit	A_1	A_2	A_3	A_4
Consumer price	Dfl				
Fuel consumption	km/l				
Maintenance, insurance	Dfl/year				
Maximum speed	km/h				
Acceleration, 0 - 100 km/h	sec				
Noise and vibrations	verbal				
Reliability	%				
Cargo volume	dm^3				
Comfort	verbal				
Ambiance	verbal				

Let us not underestimate the importance of the tableau. In many situations, once the data are on the table, the preferred alternative clearly emerges and the decision problem can easily be solved. For some consulting firms the performance tableau, which shows the comparative strengths and weaknesses of the alternatives, is accordingly the end of the analysis. The RAND Corporation, entrusted with many large water-management projects in The Netherlands such as the evaluation of the river dike improvements of 1993/94, usually presents the results of a policy analysis on a score card, which is in fact a performance tableau. Thereafter, it is left to the decision makers to arrive at a compromise solution. A simple score card is shown in Figure 2.1. The RAND Corporation explicitly states that it is sometimes easier for a group of decision makers to agree on the preferred alternative, perhaps for different reasons, than to agree on the weights to assign to the various criteria (Walker et. al., 1994). Thus, a compromise solution is supposed to emerge as the result of some magic in the decision process. This happens in many cases indeed, possibly even in the majority of the cases. Just the careful collection of data is enough to generate consensus. Miracles do not always happen, however, and that is the reason why we continue the analysis.

Impact Categories	Car	Train	Plane
Total travel time (hr)	7	9	3
Total travel cost	11	21	33
Time under stress (difficult driving, carrying bags, fighting for taxi) (hr)	4	1	2
Degree of privacy possible	Much	Some	Little

Rankings: Best, Intermediate, Worst

Figure 2.1 *Score card which shows the performance of three alternative means of transportation from Amsterdam to Berlin.*

The number of criteria in Table 2.1 is already quite large, and they are not independent. The consumer price, the fuel consumption, and the expenditures for maintenance and insurance are closely related. One can take the estimated annual expenditures (based on the estimated number of kilometers per year) or just the consumer price in order to measure how well the objective of cost minimization has been satisfied. Similarly, a high maximum speed, a rapid acceleration, and the absence of noise and vibrations contribute to the pleasure of driving a car. That pleasure may be the real criterion. The performance tableau could accordingly be reduced, but the performance indicators in it provide valuable information. They help the decision makers to remain down to earth, and they prevent that the decision makers are swept away by a nice car-body design, for instance.

Finally, the decision makers have to convert the data of the performance tableau into subjective values expressing how well the alternatives satisfy the objectives such as cost minimization and pleasure maximization. For the qualitative criteria they usually have an arithmetic scale only to express their assessment of the performance. The seven-point scale 1,...,7 which is well-known in the behavioural sciences, and the seven-point scale 4,...,10 which can easily be used for the same purposes, will extensively be discussed in the sections to follow. Under the quantitative criteria the conversion is also non-trivial. We will propose a simple and straightforward conversion procedure in the next section. The justification of the procedure requires many arguments from the behavioural sciences and from psycho-physics.

2.2 CATEGORIZATION OF A RANGE

Let us consider the subjective evaluation of cars under a number of quantitative criteria: first under the consumer-price criterion, thereafter under the reliability and the maximum-speed criterion. This will enable us to illustrate the subdivision of the ranges of acceptable performance data and the generation of judgemental categories (price categories,...). For the time being we consider the problem from the viewpoint of a single decision maker only. The subjective evaluation under qualitative criteria is also postponed to later sections and to the next chapter.

Cars under the consumer-price criterion. Usually, low costs are important for a decision maker so that he/she carefully considers the consumer price (the suggested retail price or sticker price in the USA) and possibly the annual expenditures for maintenance and insurance. The consumer price as such, however, cannot tell us whether a given car would be more or less acceptable. That depends on the context of the decision problem, that is, on the spending power of the decision maker and on the alternative cars which he/she seriously has in mind. In what follows we shall be assuming that the acceptable prices are anchored between a minimum price P_{min} to be paid anyway for the type or class of cars which the decision maker seriously considers and a maximum price P_{max} which he/she cannot or does not really want to exceed. Furthermore, we assume that the decision maker will intuitively subdivide the price range (P_{min}, P_{max}) into a number of subintervals which are felt to be subjectively equal. We take the gridpoints P_{min}, $P_{min} + e_0$, $P_{min} + e_1$, ... to denote the price levels which demarcate these subintervals. The price increments e_0, e_1, e_2,... represent the echelons of the so-called category scale under construction. In order to model the requirement that the subintervals must subjectively be equal we recall Weber's psycho-physical law of 1834 stating that the just noticeable difference Δs in stimulus intensity must be proportional to the actual stimulus intensity itself. The just noticeable difference is the smallest possible step when we move from P_{min} to P_{max}. We assume that it is practically the step carried out in the construction of our model. Thus, taking the price increment above P_{min} as the stimulus intensity, i.e. assuming that the decision maker is not really sensitive to the price as such but to the excess above the minimum price to be paid anyway for the cars under consideration, we set

$$e_\nu - e_{\nu-1} = \varepsilon\, e_{\nu-1},\ \nu = 1, 2, \ldots,$$

which yields

$$e_\nu = (1 + \varepsilon)\, e_{\nu-1} = (1 + \varepsilon)^2\, e_{\nu-2} = \ldots.. = (1 + \varepsilon)^\nu\, e_0.$$

Obviously, the echelons constitute a sequence with geometric progression. The initial step is e_0 and $(1 + \varepsilon)$ is the progression factor. The integer-valued parameter ν is chosen to designate the order of magnitude of the echelons.

The number of subintervals is small because human beings have a limited vocabulary, in the sense that they have only a small number of verbal terms or labels in order to categorize the prices. The following qualifications are commonly used as category labels:

cheap,
cheap/somewhat more expensive,
somewhat more expensive,
somewhat more/more expensive,
more expensive,
more/much more expensive,
much more expensive.

Thus, we have four major, linguistically distinct categories: *cheap, somewhat more, more*, and *much more expensive*. Moreover, there are three so-called threshold categories between them which can be used when the decision maker hesitates between the neighbouring qualifications. Let us now try to link the price categories with the price levels $P_{min} + e_0$, $P_{min} + e_1$,.....

The next section will show that human beings follow a uniform pattern in many unrelated areas when they subdivide a particular range into subjectively equal subintervals. They demarcate the subintervals by a geometric sequence of six to nine grid points corresponding to major and threshold echelons, and the progression factor is roughly 2. Sometimes there is a geometric sequence with gridpoints corresponding to major echelons only, and the progression factor is roughly 4. In the present section we use these observations in advance, in order to complete the subdivision. Let us take, for instance, the range between Dfl 20,000 (ECU 9,000, US$ 12,000) and Dfl 40,000 (ECU 18,000, US$ 24,000) for compact to mid-size cars in The Netherlands. The length of the range is Dfl 20,000. Hence, setting the price level $P_{min} + e_6$ at P_{max} we have

$$e_6 = P_{max} - P_{min},$$

$$e_0(1 + \varepsilon)^6 = 20{,}000 \text{ and } (1 + \varepsilon) = 2,$$

$$e_0 = 20{,}000/64 \approx 300.$$

Now, we associate the price levels with the price categories as follows:

$P_0 = P_{min} + e_0$	Dfl 20,300	cheap,
$P_1 = P_{min} + e_1$	Dfl 20,600	cheap/somewhat more expensive,
$P_2 = P_{min} + e_2$	Dfl 21,200	somewhat more expensive,
$P_3 = P_{min} + e_3$	Dfl 22,500	somewhat more/more expensive,
$P_4 = P_{min} + e_4$	Dfl 25,000	more expensive,
$P_5 = P_{min} + e_5$	Dfl 30,000	more/much more expensive,
$P_6 = P_{min} + e_6$	Dfl 40,000	much more expensive.

Thus, we have "covered" the price range (P_{min}, P_{max}) by the grid with the geometric sequence of points

$$P_\nu = P_{min} + (P_{max} - P_{min}) \times \frac{2^\nu}{64}, \quad \nu = 0, 1, \ldots, 6. \tag{2.1}$$

In what follows we take P_ν to stand for the ν-th price category and the integer-valued parameter ν for its order of magnitude, which is given by

$$\nu = {}^2\log\left(\frac{P_\nu - P_{min}}{P_{max} - P_{min}} \times 64\right). \tag{2.2}$$

Categorization of the prices means that each price in or slightly outside the range (P_{min}, P_{max}) is supposed to "belong" to a particular category, namely the category represented by the nearest P_ν. We refer to the cars of the category P_0 as the cheap ones within the given context, and to the cars of the categories P_2, P_4, and P_6 as the *somewhat more*, *more*, and *much more* expensive ones. At the odd-numbered grid points P_1, P_3, and P_5 the decision maker hesitates between two adjacent gradations of expensiveness. If necessary, we can also introduce the category P_8 of *vastly more* expensive cars which are situated beyond the range, as well as the category P_7 if the decision maker hesitates between *much more* and *vastly more* expensiveness. The even-numbered grid points are the so-called major grid points designating the major gradations of expensiveness. They constitute a geometric sequence in the range (P_{min}, P_{max}) with progression factor 4. If we also take into account the odd-numbered grid points corresponding to hesitations, we have a geometric sequence of major and threshold gradations with progression factor 2.

The crucial assumption here is that the decision maker considers the prices from the so-called desired target P_{min} at the lower end of the range of acceptable prices. From this viewpoint he/she looks at less favourable alternatives. That is the reason why the above categorization, in principle an asymmetric subdivision of the range under consideration, has an orientation from the lower end. The upward direction is typically the line of sight of the decision maker under the price criterion. Figure 2.2 shows the concave form of the relationship between the echelons on the interval (P_{min}, P_{max}) and their order of magnitude.

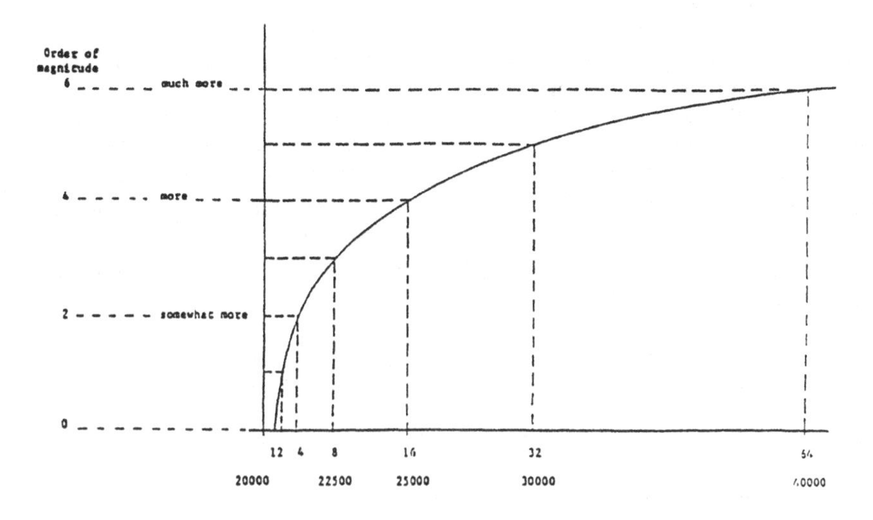

Figure 2.2 *Categorization of a price range. Concave relationship between the price echelons and their orders of magnitude.*

Suppose that the prices of the cars A_j and A_k belong to the categories represented by

$$P_{\nu_j} \text{ and } P_{\nu_k}$$

respectively. We express the relative preference for A_j with respect to A_k by the inverse ratio of the price increments above $P_{\min}$ so that it can be written as

$$\frac{P_{\nu_k} - P_{\min}}{P_{\nu_j} - P_{\min}} = 2^{\nu_k - \nu_j}. \tag{2.3}$$

By this definition, a car in the price category P_0 is 4 times more desirable than a car in the category P_2. The first-named car is said to be somewhat cheaper, the last-named car is somewhat more expensive. Hence, assuming that we also have a limited number of labels to express relative preference in comparative judgement, we identify the ratio 4:1 with weak preference. Similarly, we identify the ratio 16:1 with definite preference (the first-named car is cheaper, the last-named car more expensive), and a ratio of 64:1 with strong preference (the first-named car is much cheaper, the last-named car much more expensive). The relative preference depends strongly on $P_{\min}$ and weakly on $P_{\max}$. When

P_{max} increases, two prices which initially belong to different price categories will tend to belong to the same one.

Cars under the reliability criterion. Numerical data to estimate the reliability of cars are usually available. Consumer organizations collect information about many types and models which follow the prescribed maintenance procedures, and they publish the percentage of cars suffering from technical failures, engine malfunctions, etc., in the first three or five years (see the Annual Auto Issue of the Consumer Reports, US Consumer Union). Let us suppose that the decision maker only considers cars with a reliability of at least R_{min} = 95%, so that he/she is restricted to the interval (R_{min}, R_{max}) with R_{max} usually set to 100%. Following the mode of operation just described, we obtain the major grid points (the major categories of reliability)

$R_0 = R_{max} - e_0$	99.9%	reliable,
$R_2 = R_{max} - e_2$	99.7%	somewhat less reliable,
$R_4 = R_{max} - e_4$	98.7%	less reliable,
$R_6 = R_{max} - e_6$	95.0%	much less reliable,

because $e_0 = (100 - 95)/64 \approx 0.08$. In general we can write

$$R_\nu = R_{max} - (R_{max} - R_{min}) \times \frac{2^\nu}{64}, \quad \nu = 0, 1, \ldots, 6.$$

The alternatives are compared with respect to the desired target, which is here taken to be at the upper end R_{max} of the range of acceptable reliabilities. The relative preference is inversely proportional to the distance from the target. If we take the symbols

$$R_{\nu_j} \text{ and } R_{\nu_k}$$

to denote the reliability of the alternative cars A_j and A_k respectively, then the inverse ratio

$$\frac{R_{max} - R_{\nu_k}}{R_{max} - R_{\nu_j}} = 2^{\nu_k - \nu_j}$$

represents the relative preference for A_j with respect to A_k under the reliability criterion. The qualification "somewhat more reliable" implies that the inverse ratio of the distances to the respective target is 4:1, the qualification "more reliable" implies that the inverse ratio is 16:1, etc. The relationship between the order of magnitude ν and the reliability category R_ν takes the explicit form

$$\nu = {}^2\log\left(\frac{R_{max} - R_\nu}{R_{max} - R_{min}} \times 64\right). \tag{2.4}$$

The typical relationship between the echelons on the dimension of reliability and the orders of magnitude is shown in Figure 2.3.

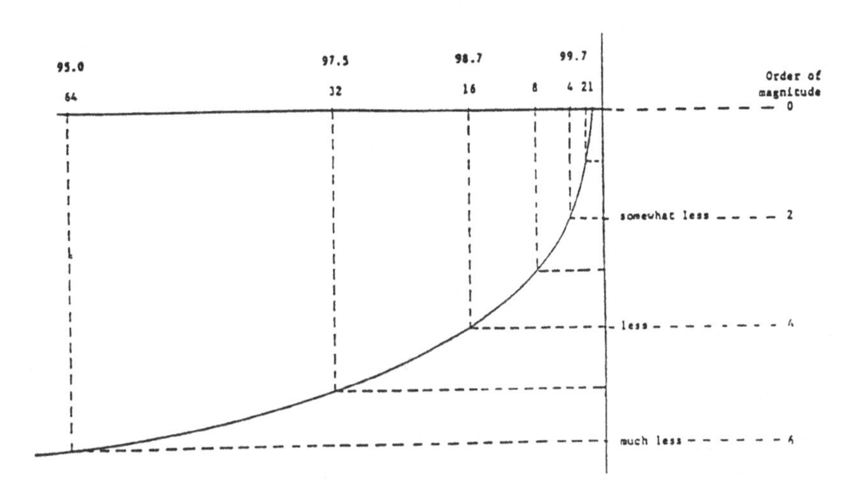

Figure 2.3 *Categorization of a reliability range. Concave relationship between the echelons on the dimension of reliability and their orders of magnutude.*

Cars under the maximum-speed criterion. It may happen that the categorization starts, not from the desired target at one end of the range, but from the opposite end point because the desired target is hazy. An example is given by the categorization of the maximum velocities. For many drivers in our country, where the maximum speed on the motor ways is 120 km/h, the range of acceptable maximum velocities has a clear lower end point M_{min} at 140 km/h. Even if the drivers consistently prefer higher maxima to lower ones, the desired target is difficult to specify. Let us set M_{max} at 220 km/h (a very high speed which is only occasionally possible on the German motor ways). It seems to be reasonable to choose the orientation from M_{min} so that we have the following major grid points:

$M_0 = M_{min} + e_0$	141 km/h	slow,
$M_2 = M_{min} + e_2$	145 km/h	somewhat faster,
$M_4 = M_{min} + e_4$	160 km/h	faster,
$M_6 = M_{min} + e_6$	220 km/h	much faster.

In general we have

$$M_\nu = M_{min} + (M_{max} - M_{min}) \times \frac{2^\nu}{64}, \nu = 0,1,\ldots,6,$$

where the order of magnitude v and the category M_v are connected by the relation

$$v = {}^2\log\left(\frac{M_v - M_{\min}}{M_{\max} - M_{\min}} \times 64\right). \tag{2.5}$$

The relative maximum speed of two alternative cars A_j and A_k with maximum speeds

$$M_{v_j} \text{ and } M_{v_k}$$

is given by the ratio

$$\frac{e_{v_j}}{e_{v_k}} = \frac{M_{v_j} - M_{\min}}{M_{v_k} - M_{\min}} = 2^{v_j - v_k},$$

not by the inverse ratio of the echelons as in the previous cases. The choice of the orientation is left to the decision maker. What matters is his/her perspective on the decision problem.

Desired targets in the interior of the range. When the decision maker chooses the desired target in the interior of the range of acceptable performance data, it usually means that he/she tries to satisfy a hidden objective. When he/she prefers a maximum speed of 180 km/h over 220 km/h, for instance, then he/she is possibly taking into account the higher risk associated with high-speed driving. We mostly suggest the decision maker to consider the minimization of risk or the maximization of safety separately, under another criterion. The decision maker may also use the proposal of Van den Honert and Lootsma (1999) to categorize the range from an interior target or reference point.

Dependence on the units of measurement. The subdivision of the range of acceptable performance data introduced in the present section has the additional advantage that the judgemental categories do not really depend on the units of measurement. Let us illustrate this under the fuel-consumption criterion. The decision maker may evaluate the fuel consumption of the alternative cars on the range between 10 and 14 km/liter or (more or less equivalently) between 7 and 10 liter/100 km. If he considers the respective ranges from the corresponding targets (maximum of 14 km/liter and minimum of 7 liters/100 km), there may be some discrepancies in his judgement but the effects are negligible.

Value functions versus categories. Figure 2.1 and Figure 2.2 suggest that we are working towards the introduction of a value function whereby to each element in the range of acceptable performance data a value between zero and one is assigned, the degree to which the objective of the underlying criterion has been satisfied (Multi-Attribute Value Theory, see Von Winterfeldt and Edwards, 1986). Indeed, one could in principle do so. On the basis of the considerations in the present section and in the next one, however, the graph of a value function should have the form of a staircase. Such a function has several discontinuities so that it is not easy to handle in a mathematical analysis. Hence, we have chosen to categorize the range of acceptable performance data and to use a small number of categories only. This is a well-known cognitive process (Rosch, 1978). A category is clearly a set of elements which are roughly equivalent for the purposes at hand. It is usually defined, not by its boundary or by an exhaustive list, but by one or more highly prototypical elements. Hence, one can deal with a category on the basis of a prototype (here the corresponding gridpoint or price level), even in the absence of precise information about its boundaries. The formation of categories is based upon two general principles. First, cognitive economy leads to a system of categories which provides a great deal of information with a reduced cognitive effort. The human observer reduces an infinite number of possible elements to a small number of collections of practically equivalent elements. Second, such a system is usually designed to map the perceived structure in the surrounding world. If there are many elements with the same combination of properties, if certain pairs, triples,... of properties appear frequently, human beings usually generate categories to structure their observations. Several MCDA methods have been extended to model judgemental categories via fuzzy concepts like fuzzy numbers. For more information the reader is referred to the author (1997).

2.3 CATEGORY SCALING IN VARIOUS AREAS

It is surprising to see how consistently human beings categorize certain intervals of interest in totally unrelated areas. In this section we present some examples to show, for instance, how they partition certain ranges on the dimensions of time and space, and how they categorize light and sound intensities.

Historical periods. The written history of Europe from 3000 BC until today is subdivided into a small number of major periods. Looking backwards from 1989, the year when the Berlin Wall was opened, one finds the following turning points, each marking off the start of a characteristic development:

1947	42 years before 1989	beginning of cold war and decolonization,
1815	170 years before 1989	beginning of the West-European dominance,
1500	500 years before 1989	beginning of world-wide trade,
450	1550 years before 1989	beginning of the Middle Ages,
-3000	5000 years before 1989	beginning of the ancient history.

These major echelons, measured by the number of years before 1989, constitute a geometric sequence with the progression factor 3.3. One obtains a more refined subdivision by the introduction of the years:

1914	75 years before 1989	beginning of the world-wars period,
1700	300 years before 1989	beginning of modern science,
1100	900 years before 1989	beginning of the High Middle Ages,
-800	2800 years before 1989	beginning of the Greek/Roman history.

With these turning points interpolated between the major ones we obtain a geometric sequence with the progression factor 1.8.

Planning horizons. In industrial planning one usually observes a hierarchy of planning cycles where decisions under higher degrees of uncertainty and with more important consequences for the company are prepared at increasingly higher management levels. The planning horizons constitute a geometric sequence, as the following list readily shows:

1 week		weekly production scheduling,
1 month	4 weeks	monthly production scheduling,
4 months	16 weeks	ABC planning of labour and tools,
1 year	52 weeks	capacity adjustment,
4 years	200 weeks	production allocation,
10 years	500 weeks	strategic planning of company structure.

The progression factor of these major echelons is 3.5. In practice, there are no planning horizons between the major ones.

Size of nations. The above categorization is not only found on the time axis but also in spatial dimensions, when the nations are categorized on the basis of population size. The major echelons in the list to follow reveal a somewhat European bias. Omitting very small nations with less than 1 million inhabitants (such as Andorra, Liechtenstein, Monaco, and Luxemburg), one has:

small nations	4 million	Denmark, Norway, Greece,....
medium-size nations	15 million	The Netherlands,....
large nations	60 million	France, Germany, Italy, United Kingdom,....
very large nations	240 million	Russia, USA,...
giant nations	1000 million	India, China.

We find again a geometric sequence, now with the progression factor 4. It seems to be reasonable to interpolate the threshold echelons:

small/medium-size	8 million	Austria, Belgium, Hungary, Sweden,....
medium-size/large	30 million	Poland, Spain,.....
large/very large nations	120 million	Japan.

The refined sequence of echelons has the progression factor 2.

Loudness of sounds. Vigorous research in psycho-physics has revealed that there is a functional relationship between the intensity of physical stimuli (sound, light,....) on the one hand and the sensory responses (the subjective estimates of the intensity) on the other. Psycho-physics starts from Weber's law of 1834 stating that the just noticeable difference Δs in stimulus intensity must be proportional to the actual stimulus level s itself. In Fechner's law of 1860, the sensory response $\Delta\Psi$ to a just noticeable difference Δs is supposed to be constant, which implies that $\Delta\Psi$ would be proportional to $\Delta s/s$. Integration yields a logarithmic relationship between Ψ and s. Thus, with s_1 and s_2 representing intensity levels of a particular stimulus such as sound or light, the sensory and the physical intensities would be connected by

$$\Psi(s_1) - \Psi(s_2) = \alpha \ln (s_1/s_2),$$

where α stands for an unknown positive constant. Obviously, a geometric sequence of stimulus intensities would yield an arithmetic sequence of sensory responses.

Additional experience has finally shown that Fechner's law does not hold. Brentano suggested in 1874 that the sensory response $\Delta\Psi$ to a just noticeable difference might be proportional to the response level Ψ, so that $\Delta\Psi/\Psi$ would be proportional to $\Delta s/s$. By integration one obtains that Ψ would be a power function of s. Empirical evidence in many areas of sensory perception prompted Stevens (1957) eventually to postulate the power law as a general psycho-physical law. Thus, the relationship between the sensory and the physical intensity ratios can be written as

$$\Psi(s_1)/\Psi(s_2) = (s_1/s_2)^{\beta}.$$

The positive exponent β has been established for many sensory systems under precisely defined circumstances. For a 1000 Hertz tone it is roughly 0.3. The power law clearly implies that a geometric sequence of stimulus intensities yields a geometric sequence of sensory responses, albeit with a different progression factor.

It is customary in acoustics to use a decibel scale for sound intensities. Thus, the intensity s with respect to a reference intensity s_0 is represented by

$$\text{dB}(s) = 10 \times {}^{10}\log(s/s_0).$$

A difference of 10 dB between sound intensities s_1 and s_2 can henceforth be written as

$$\text{dB}(s_1) - \text{dB}(s_2) = 10 \times {}^{10}\log(s_1/s_2) = 10,$$

which implies

$$s_1/s_2 = 10,$$
$$\Psi(s_1)/\Psi(s_2) = (s_1/s_2)^{\beta} \approx 2.$$

In other words, by an additive step of 10 dB the sound intensity is felt to be doubled. In summary, an arithmetic sequence on the dB-scale with steps of 10 dB represents a geometric sequence of physical sound intensities with the progression factor 10, and this is felt to be a geometric sequence with the progression factor 2. The interesting result for our purposes is that the range of audible sounds has roughly been categorized as follows:

40 dB	very quiet, whispering,
60 dB	quiet, conversation,

80 dB	moderately loud, electric mowers, foodblenders,
100 dB	very loud, farm tractors, motorcycles,
120 dB	painfully loud, jets during take-off.

Although the precision should be taken with a grain of salt because there is a mixture of sound frequencies at each of these major echelons, we obviously find here a geometric sequence of subjective sound intensities with the progression factor 4.

Brightness of light. Physically, the perception of light and the perception of sound proceed in different ways, but these sensory systems follow the power law with practically the same value of the exponent β. Hence, a step of 10 dB in light intensity is felt to double the subjective brightness. The range of visible light intensities has roughly been categorized as follows:

40 dB	star light,
60 dB	full moon,
80 dB	street lighting,
100 dB	office space lighting,
120 dB	sunlight in summer.

Under the precaution that the precision should not be taken too seriously because there is a mixture of wavelengths at each of these major echelons, we observe that the subjective light intensities also constitute a geometric sequence with the progression factor 4.

For a more detailed documentation on psycho-physics we refer the reader to Marks (1974), Michon et al. (1978), Roberts (1979), Zwicker (1982), and Stevens and Hallowell Davis (1983). The reader will find that the sensory systems for the perception of tastes, smells, and touches follow the power law with exponents in the neighbourhood of one. We did not see a categorization such as we found for loudness and brightness, so that we have neither additional evidence nor counter-evidence for the geometric progression described in the above examples.

2.4 ASSESSING THE ALTERNATIVES, DIRECT RATING

When decision makers judge the performance of the alternatives, they frequently express their judgement by choosing an appropriate value between a predetermined lower limit for the worst (real or imaginary) alternative and a predetermined upper limit for the best (real or ideal) alternative. In schools and universities this direct-rating procedure is known as the assignment of grades expressing the performance of the pupils or students on a category scale with equidistant steps, between 1 and 5, between 1 and 10, or between 1 and 100. The upper limit varies from country to country. Sometimes the scale is upside down so that the grade 1 is used to express excellent performance. Because everybody has once been subject to his/her teacher's judgement the grades have a strong qualitative connotation which can successfully be used in MCDA. Concentrating on the scale between 1 and 10 we suppose that a unit step difference represents an order of magnitude difference in performance. A pupil who usually or on average scores 9 is an order of magnitude better than a pupil scoring 8, etc. A unit step difference designates a performance ratio 2.

Direct rating under quantitative criteria. Returning to the car selection problem we accordingly assign the following grades:

10	excellent	order of magnitude $\nu = 0$,
8	good	order of magnitude $\nu = 2$,
6	fair	order of magnitude $\nu = 4$,
4	unsatisfactory	order of magnitude $\nu = 6$,

to the major gradations of expensiveness and reliability. In pass-or-fail decisions at schools the grades 1, 2, and 3 are normally used for a (very) poor performance that cannot be compensated elsewhere so that we mostly ignore them. We will mainly use the seven-point scale 4, ..., 10, and occasionally the nine-point scale 2, 3, ..., 10. Now, considering two alternative cars A_j and A_k under the consumer-price criterion with the respective grades

$$g_j = 10 - \nu_j \text{ and } g_k = 10 - \nu_k \tag{2.6}$$

assigned to them, we take the inverse ratio

$$\frac{e_{\nu_k}}{e_{\nu_j}} = \frac{P_{\nu_k} - P_{\min}}{P_{\nu_j} - P_{\min}} = 2^{g_j - g_k}$$

to stand for their relative expensiveness. The relative reliability of the two alternatives is scored in a similar way. Figure 2.4 and Figure 2.5 illustrate the relationship between judgemental categories, orders of magnitude, and grades.

For maximum speed we have a somewhat different assignment of grades. We take

$$g_j = 4 + \nu_j \text{ and } g_k = 4 + \nu_k \tag{2.7}$$

since we do not have an orientation from the desired target but from the opposite end of the range. Thus, the relative maximum speed of the two alternatives is expressed by the ratio

$$\frac{e_{\nu_j}}{e_{\nu_k}} = \frac{M_{\nu_j} - M_{\min}}{M_{\nu_k} - M_{\min}} = 2^{g_j - g_k},$$

not by the inverse ratio of maximum velocities above $M_{\min}$.

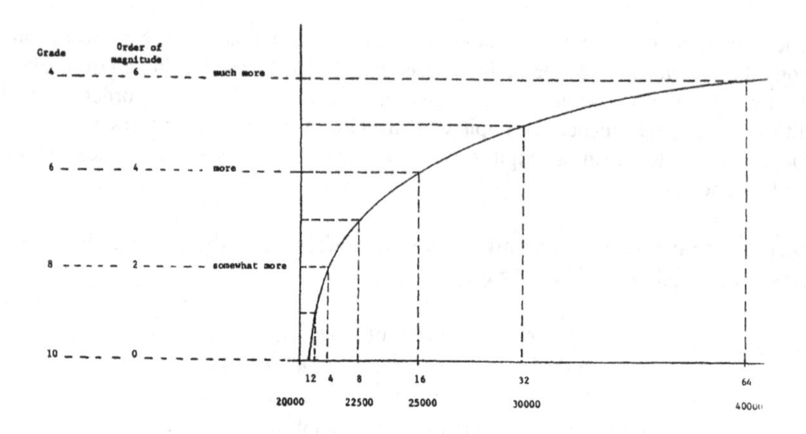

Figure 2.4 *Categorization of a price range. The geometric progression of the echelons in the horizontal direction is obtained by a series of bisections or duplications. In the vertical direction one counts the number of steps.*

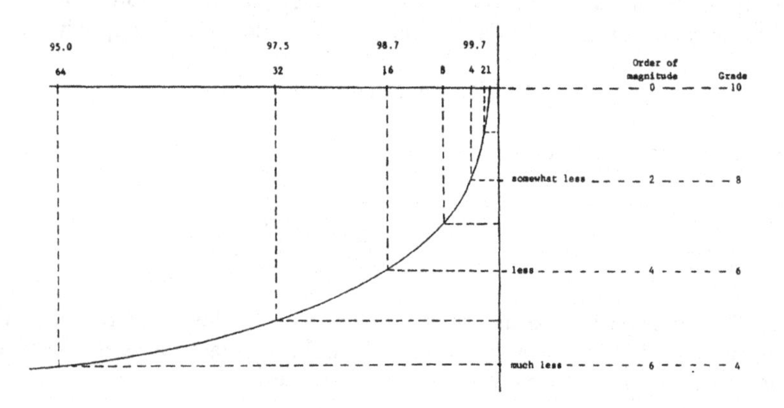

Figure 2.5 *Categorization of a reliability range. The geometric progression of the echelons in the horizontal direction is obtained by a series of bisections or duplications. In the vertical direction one counts the number of steps.*

Ratio and difference judgement. In the psycho-physical literature the issue of how human beings judge the relationship between two stimuli on one single dimension was brought up a few decades ago. Torgerson (1961) observed that they perceive only one quantitative relation, but they estimate differences in subjective values when they are requested to express their judgement on a scale with arithmetic progression and they estimate ratios when the proposed scale is geometric. Thus, they interpret the relationship as it is required in the experiment. Which of the two interpretations is correct cannot empirically be decided because these are alternative ways of saying the same thing. Moreover, difference judgement is logarithmically related to ratio judgement. Psychophysical research in the seventies and eighties (Veit, 1978; Birnbaum, 1982) confirmed Torgerson's observation that comparative judgement of two stimuli uses one operation only. This is all clearly illustrated by the measurement of sound and light intensities, where a difference on the decibel scale represents the logarithm of an intensity ratio (logarithmic coding, see Bruce and Green, 1990). Moreover, in the assessment of pupils and students at schools and universities, when we move upwards from one grade to the next, the corresponding performance seems to increase exponentially.

Because we work in fact with differences of grades only, there is an additive degree of freedom in the grades which enables us to replace the seven-point scale 4,...., 10 by the seven-point scale 1,..., 7 which is well-known in the behavioural sciences. Similarly, we can convert the symbolic seven-point scale ranging from - - - to + + + into the numerical seven-point scale 4,..., 10. Thus, the decision maker has a variety of scales to express his judgement, but there is a uniform approach to analyze the responses. This can successfully be used in planning (Van Gennip et al., 1997).

Let us illustrate the direct-rating procedure via the car selection problem. Under the assumption that the decision maker considers prices in the range between Dfl 20,000 and Dfl 40,000, reliabilities between 95% and 100%, and maximum velocities between 140 km/h and 220 km/h (by these ranges we model the context of the problem), we obtain the correspondence between objective data and subjective judgement as shown in Table 2.2.

In general, the assignment of judgemental data (the qualifications or grades in the columns 4 - 6 of Table 2.2) does not depend on the units of measurement. If the fuel consumption of a car is low in km/liter, for instance, it is also low in miles/gallon or in liters/100 km. We leave it to the reader to categorize the equivalent ranges between 10 and 14 km/liter, between 24 and 33 miles/gallon, and between 7 and 10 liters/100 km.

Table 2.2 *Assignment of grades within predetermined ranges. The decision maker may use a number of seven-point scales: a verbal scale between unsatisfactory and excellent, a numerical scale with the grades* 4, ..., 10, *and a symbolic scale between* - - - *and* + + +.

Price	Reliability	Max. speed	Performance	Grade	Symbolic scale
20,300	99.9	220	Excellent	10	+++
20,600	99.8	180	Good/excellent	9	++
21,200	99.7	160	Good	8	+
22,500	99.2	150	Fair/good	7	o
25,000	98.7	145	Fair	6	-
30,000	97.5	143	Unsatisfactory/fair	5	- -
40,000	95.0	141	Unsatisfactory	4	- - -

Direct rating under qualitative criteria. The assignment of grades which we have just sketched is feasible when the performance of the alternatives can be expressed in physical or monetary units on a one-dimensional scale. We also use a direct-rating procedure, however, when the performance can only be expressed in verbal terms. In the evaluation of cars under the criterion of car-body design, for instance, we may ask the decision makers to rate the cars straightaway. First, they have to determine the endpoints in order to frame the analysis. They are requested to imagine the worst and the best (real or imaginary) alternative and to assign proper qualifications and/or grades to them. Thereafter, they may carefully describe various intermediate performance levels. This will enable them to interpolate the actual decision alternatives between the endpoints. It is not always an easy procedure but since it is current practice in schools and universities we assume that it is workable, at least for somewhat experienced decision makers. Under the criterion of car-body design, for instance, the decision makers may draw up the verbal scale of Table 2.3.

Table 2.3 *Verbal scale for the direct rating of cars under the criterion of car-body design. Grades can be assigned to the qualifications in the usual manner (excellent* = 10,....., *unsatisfactory* = 4, *for instance) depending on the decision maker's judgement.*

Excellent	An innovative, exciting, trend-setting design. Examples in the history of the motor car: the Austin Mini, the Citroen DS, the Renault 16, and the Renault Espace.
Good	A modern, well-balanced body, suggesting that the owner can afford an easy-going life-style, both during the week and in the weekend. This seems to apply to many wagons and multi-purpose vehicles.
Fair	A conventional wind-tunnel design, not very appealing, not very interesting, but the car is intended to be a reliable work-horse only.
Unsatisfactory	Despite their provocative ugliness, they were on the market for three or four decades: the Citroen 2CV and the Volkswagen Beetle.

2.5 CRITERION WEIGHTS AND AGGREGATION

Let us first introduce some notation and terminology. We consider a finite number of alternatives A_j, $j = 1,....., n$, under a finite number of performance criteria C_i, $i = 1,..., m$, with the respective criterion weights c_i, $i = 1,...., m$. Furthermore, we assume that the criterion weights are normalized so that they sum up to 1. The decision maker assessed the alternatives under each of the criteria separately. Moreover, the decision maker expressed his/her judgement of alternative A_j under criterion C_i by the assignment of the grade g_{ij} which from now will be called the impact grade. So far, we have been working on different dimensions such as consumer price, reliability, and maximum speed. Judgemental statements like "somewhat more expensive" and "somewhat more reliable" cannot be aggregated, however, unless we make a transition to the new, common dimension of desirability or preference intensity. That is the reason why we take the expression "somewhat more reliable", for instance, to stand for "somewhat more desirable" or "weakly preferred" under the reliability criterion. Similarly, we assume that the expression "somewhat more expensive" stands for "somewhat less desirable" under the consumer price criterion, etc.

In order to judge the overall performance of the alternatives under all criteria simultaneously we calculate the final grades f_j of the respective alternatives A_j according to the so-called arithmetic-mean aggregation rule

$$f_j = \sum_{i=1}^{m} c_i g_{ij}, \; j = 1,..., n. \tag{2.8}$$

The highest final grade is supposed to designate the preferred alternative. Let us first discuss the significance of the criterion weights, however, as well as their elicitation.

We start from the assumption that criteria have particular values or weights in the mind of the decision maker. That is not immediately clear, because one might argue that the values or weights are relevant in a particular situation only. They could depend on the manner in which the performance of the alternatives under each of the criteria individually has been recorded, that is, on the units of performance measurement. They could also depend on the aggregation procedure generating the final grades which express the performance of the alternatives under all criteria simultaneously. Usually, however, decision makers ignore these issues (Vincke, 1989; Mousseau, 1995). They are prepared to estimate the weights of the criteria or their relative importance (weight ratios), regardless of how the performance of the alternatives has been measured and regardless of the aggregation procedure, so that they seem to supply meaningless information. The underlying reason may be that the criteria have emotional or social values which neither depend on the actual decision problem itself nor on the method of analysis. Many decision makers want to be consistent over a coherent collection of decision problems. They want to employ a uniform set of criterion weights, to be used in a certain area (water management, environmental protection, national energy planning, ...) but in isolation from immediate context. Moreover, there is a good deal of distributed decision making in large organizations. The evaluation of a number of decision alternatives is entrusted to a committee, the criteria are suggested in vague verbal terms or firmly prescribed by those who established the committee, but the choice of the criterion weights and the final aggregation are felt to be the responsibility of administrators or politicians at higher levels in the hierarchy.

In SMART the decision maker may ignore the units of performance measurement because the grades do not depend on them. Under the consumer price criterion, for instance, he/she may replace Dutch florins by US dollars or any other currency, but this does not affect the orders of magnitude of the price categories. The author (1993, 1996, see also Section 3.3) presented a more detailed discussion of the issue in a study on SMART and the Multiplicative AHP, a multiplicative variant of the Analytic Hierarchy Process (Saaty, 1980). The relative importance of the criteria appeared to be a meaningful concept, even in isolation from immediate context. In the present chapter we shall be working with the arithmetic-mean aggregation rule without further discussion.

We now consider the numerical scale to quantify the relative importance (the weight ratio) of any two criteria. The first thing we want to establish is the range of possible values for the relative importance. Equal importance of two criteria is expressed by the ratio 1:1 of the criterion weights, but how do we express *much more* or *vastly more* importance? By the ratio 100:1 of the criterion weights? Is that reasonable or is that just a wild guess? In order to answer the question we carry out an imaginary experiment: we ask the decision maker to consider two real or imaginary alternatives A_j and A_k and two criteria such that his/her preference for A_j over A_k under the first criterion C_f is roughly equal to his/her preference for A_k over A_j under the second criterion C_s. Moreover, we suppose that the situation is extreme: the impact grades assigned to the two alternatives are 6 units apart under each of the two criteria. That is the maximum distance on seven-point scales like 4,....,10 and 1,...,7. We can accordingly write the impact grades assigned to A_j as

$$g_{fj} \text{ and } g_{sj},$$

and the impact grades assigned to A_k as

$$g_{fk} = g_{fj} - 6 \text{ and } g_{sk} = g_{sj} + 6.$$

This means that the decision maker has a strong preference for A_j over A_k under the first criterion C_f and an equally strong but opposite preference under the second criterion C_s (substitute $g_{fj} = 10$ and $g_{sj} = 4$, or $g_{fj} = 7$ and $g_{sj} = 1$, in order to see what happens). If the two criteria are felt to be equally important, the final grades of the two alternatives will be equal so that the decision maker is indifferent between the two alternatives under the two criteria simultaneously. However, if the final grades are 5 units apart, the ratio ω of the corresponding criterion weights has to satisfy the relation

$$\left\{\frac{\omega}{\omega+1}g_{fj} + \frac{1}{\omega+1}g_{sj}\right\} - \left\{\frac{\omega}{\omega+1}(g_{fj} - 6) + \frac{1}{\omega+1}(g_{sj} + 6)\right\} = 5, \qquad (2.9)$$

whence $\omega = 11$. Such a ratio implies that the strong preference for A_j over A_k under C_f almost completely wipes out the equally strong but opposite preference under C_s. Let us also consider the nine-point scale 2, 3,..., 10. If the impact grades of the two alternatives are 8 units apart under each of the two criteria (the maximum distance on the nine-point scale), and if the final grades are 7 units apart the ratio ω has to satisfy the relation

$$\left\{\frac{\omega}{\omega+1}g_{fj} + \frac{1}{\omega+1}g_{sj}\right\} - \left\{\frac{\omega}{\omega+1}(g_{fj} - 8) + \frac{1}{\omega+1}(g_{sj} + 8)\right\} = 7, \qquad (2.10)$$

which yields $\omega = 15$. In addition, we now introduce two new assumptions: (a) the number of gradations to express relative preference for the alternatives equals the number of gradations for the relative importance of the criteria, and (b) the numerical values associated with the gradations of relative importance constitute a sequence with geometric progression. In the extreme case of formula (2.9), where a *much higher* preference under the first criterion is practically wiped out by a *much higher* preference under the second criterion, we accordingly refer to the relative importance of the first criterion with respect to the second one as *much higher*. Similarly, in the extreme case of formula (2.10), the relative importance of the first criterion with respect to the second one is *vastly higher*. So, a ratio of 16:1 may be taken to stand for *vastly more* importance. This is also confirmed by other imaginary experiments where a decision maker is supposed to have a very strong preference for A_j under the first of three criteria and a definite preference for A_k under the second and the third criterion.

A simple geometric sequence of values, with echelons corresponding to *equal, somewhat more, more, much more*, and *vastly more* importance, and "covering" the range of values between $^1/_{16}$ and 16, is given by the sequence $^1/_{16}$, $^1/_8$, $^1/_4$, $^1/_2$, 1, 2, 4, 8, 16 with progression factor 2. Hence, we obtain the following geometric scale for the major gradations of relative importance:

16	C_f vastly more important than C_s,
8	C_f much more important than C_s,
4	C_f more important than C_s,
2	C_f somewhat more important than C_s,
1	C_f as important as C_s etc.,

and if we allow for threshold gradations to express hesitations between two adjacent qualifications in the above list, we have a geometric sequence with progression factor $\sqrt{2}$.

We can now ask the decision maker to express the importance of the criteria in grades on the scale 4, 5, ..., 10 (grades lower than 4 are possible but they practically eliminate the corresponding criteria). A difference of 6 units has to represent the ratio 8. This can be achieved via the progression factor $\sqrt{2}$. Taking h_i to stand for the grade assigned to criterion C_i, we estimate the ratio of the weights of two criteria C_f and C_s by

$$\left(\sqrt{2}\right)^{h_f - h_s}$$

An unnormalized weight of C_i is accordingly given by

$$\left(\sqrt{2}\right)^{h_i}.$$

The normalization thereafter, whereby we eliminate the additive degree of freedom in the grades, yields the desired weight

$$c_i = \frac{\left(\sqrt{2}\right)^{h_i}}{\sum_i \left(\sqrt{2}\right)^{h_i}} \tag{2.11}$$

of criterion C_i. The criterion weights clearly sum up to 1. Thus, because there is a uniform scale to judge the alternatives under the respective criteria, we can easily quantify the gradations of the relative importance of the criteria.

2.6 THE CAR SELECTION PROBLEM

Let us illustrate the results obtained so far by a concrete example. We consider a decision maker who normally drives a Ford, possibly because he/she likes the manufacturer as well as the dealer in the vicinity. Given the changing circumstances in the family, the question is to choose the appropriate size of the car. It is not necessary for the new car to be of the same size and capacity as the old one.

In the first round the decision maker concentrates on a Fiesta, an Escort, and a Mondeo (the Contour in the USA). The relevant criteria and the relevant data to be found in various publications of consumer organizations (the Yearbook 1995 of the Royal Netherlands Automobile Club KNAC and the Price List for New and Used Cars of the Automobile Association of the Netherlands ANWB) as well as the marks which the decision maker assigns to the criteria are summarized in the performance tableau of Table 2.4. For the time being he/she does not consider the reliability criterion. The cars are all made by the same manufacturer, but the historical data for the recently introduced Mondeo are not yet complete. He/she also excludes the costs for maintenance, insurance, and fuel consumption. They are related to the consumer price. Furthermore, the decision maker chooses the following ranges for his/her categorical judgement:

Consumer price	Dfl 20,000 - 40,000 (for compact to mid-size cars),
Maximum speed	140 - 220 km/h (for the European motor ways),
Acceleration	8 - 20 sec (to reach a speed of 100 km/h),
Cargo volume	200 - 2000 dm^3 (wagon with the rear seat down).

Note that the ranges cannot arbitrarily be chosen. On the contrary, they must be compatible. The decision maker may examine, for instance, what a client can reasonably expect in the price range of Dfl 20,000 - 40,000, in order to set the ranges under the remaining criteria. This step urges him/her to scan the market and to assess the alternatives within a particular framework which should not be too narrow (see the decision traps described by Russo and Schoemaker, 1989).

Table 2.4 *Performance tableau for a car selection problem. Three Ford cars with widely varying engines under four quantitative criteria selected to minimize the consumer price and to maximize the maximum speed, the acceleration, and the cargo volume.*

Criterion	Grade	Fiesta 1.3 44 kW	Escort 1.6 66 kW	Mondeo 1.8 85 kW
Consumer price	9	25,000	33,000	40,000
Maximum speed	5	153	177	195
Acceleration	7	15.3	12.3	11.1
Cargo volume	6	250	380	480

In Table 2.5 we summarize the assignment of grades to the performance data within the predetermined ranges (which do not depend on the alternatives in the actual decision problem). Note that the categorization starts from the desired target under the criteria of consumer price and acceleration, but from the opposite end under the criteria of maximum speed and cargo volume. With the data of Table 2.4 and the assignment of grades in Table 2.5 the decision maker can now easily obtain the criterion weights and the final grades of Table 2.6. The final grades normally reveal the preferred alternative but here they are so close that the decision maker seems to be indifferent between the three cars.

Table 2.5 *Assignment of grades within predetermined ranges under four quantitative criteria, the consumer price, the maximum speed, the acceleration, and the cargo volume.*

Cons. price	Max. speed	Acceleration	Cargo volume	Performance	Grade
20,300	220	8.2	2000	Excellent	10
20,600	180	8.4	1100		9
21,200	160	8.8	650	Good	8
22,500	150	9.5	425		7
25,000	145	11	310	Fair	6
30,000	143	14	255		5
40,000	141	20	230	Unsatisfactory	4

Table 2.6 *Final tableau for a car selection problem. Three Ford cars with widely varying engines under four quantitative criteria.*

Criterion	Weight	Fiesta 1.3 44 kW	Escort 1.6 66 kW	Mondeo 1.8 85 kW
Consumer price	0.475	6	4.5	4
Maximum speed	0.119	7.5	9	9.5
Acceleration	0.238	4.5	5.5	6
Cargo volume	0.168	5	6.5	7.5
Final scores		5.7	5.6	5.7

Note that the impact grades in Table 2.6 may also be obtained via the relationships between the orders of magnitude and the impact grades as given in Section 2.2. Using the formulas (2.2) and (2.6), for instance, the decision maker could derive the impact grade for the Fiesta under the consumer price criterion by the calculation of the expression

$$10 - {}^{2}\log\left(\frac{25{,}000-20{,}000}{40{,}000-20{,}000}\times 64\right) = 6.0.$$

By the formulas (2.5) and (2.7), the impact grade of the Fiesta for maximum speed follows from the evaluation of the expression

$$4 + {}^{2}\log\left(\frac{153-140}{220-140}\times 64\right) = 7.4.$$

We have rounded off the calculated results to the nearest multiple of 0.5 because the seven-point scale 4, 5, ..., 10 is sometimes felt to be rather coarse, but the effect is negligible.

The final grades in Table 2.6 may come as a surprise. In many projects we have indeed observed that the final grades produced in the first round of the analysis seem to be incredible because the decision maker is not always aware of his/her preferences. Moreover, he/she does not always realize how (un)important some of the criteria may be. The subsequent discussions and deliberations, followed by a reformulation of the criteria and a new exploration of the possible alternatives, may lead to an equilibrium after the second or possibly the third round: the decision maker agrees that the analysis brought up a compromise between his/her conflicting preferences and aspirations.

Table 2.6 possibly demystifies the selection of cars. There seems to be a fairly consistent price-performance relationship when we weigh the technical features against the consumer price only, so that the decision maker remains indifferent between the alternatives. In such a situation qualitative criteria like comfort and ambiance are usually decisive.

Let us suppose that the decision maker extends the analysis in order to include the products of other car manufacturers as well. He/she decides to concentrate on the Ford Mondeo 1.8 and on other mid-size cars like the Renault Laguna 2.0 and the Honda Accord 2.0. The relevant data are summarized in Table 2.7. We assume that the criteria and the criterion weights remain unchanged. The fact that the price of the Honda is slightly beyond the original range of acceptable prices urges the decision maker to extend the price range and to set the upper end at Dfl 45,000, for instance. Thus, throughout the analysis the range includes the alternatives which are in principle acceptable (see also the warnings of Russo and Schoemaker, 1989). With the data of Table 2.7 and the assignment of grades in Table 2.5 the decision maker obtains the final grades of Table 2.8. The final grades are again so close that the decision maker seems to be indifferent between the three mid-size cars as well.

Table 2.7 *Performance tableau for a car selection problem. Three mid-size cars of different manufacturers under four quantitative criteria.*

Criterion	Grade	Renault Laguna 2.0 83 kW	Honda Accord 2.0 96 kW	Ford Mondeo 1.8 85 kW
Consumer price	9	41,000	44,000	40,000
Maximum speed	5	200	200	195
Acceleration	7	10.6	10.2	11.1
Cargo volume	6	452	405	480

Table 2.8 *Final tableau for a car selection problem. Three mid-size cars of different manufacturers under four quantitative criteria.*

Criterion	Weight	Renault Laguna 2.0 83 kW	Honda Accord 2.0 96 kW	Ford Mondeo 1.8 85 kW
Consumer price	0.475	4.5	4.0	4.5
Maximum speed	0.119	9.5	9.5	9.5
Acceleration	0.238	6	6.5	6
Cargo volume	0.168	7	7	7.5
Final scores		6.1	6.0	6.2

The decision maker may now take into account that the Japanese car makers are consistently reported to outperform their European and American competitors with regard to the reliability of their products. The Annual Auto Issue 1996 of the Consumer Reports, US Consumer Union, reports that the Honda Accord has a predicted reliability of more than 98% over the first three years, whereas the Ford Contour has a predicted reliability between 91% and 95%. Since the cumulative data over a three years period are not readily available for the Renault the decision maker has to use his/her judgement only. Table 2.2 may be used now to assign grades under the criterion of reliability. In addition, the decision maker has to consider the ambiance of the cars. Managers in a multi-national company, for instance, are not entirely free in their choice. They create a certain amount of goodwill if they buy a car which has been designed and produced in the home country of the parent company. Let us suppose here that the decision maker in question therefore has some preference for the ambiance of American cars.

In fact, there is a mixture of quantitative and qualitative criteria now. This phenomenon, quite normal in MCDA, can easily be handled via SMART. Under the qualitative criteria one works via the direct assignment of grades, without the calculation of orders of magnitude. Qualifications like *somewhat less* and *somewhat more* constitute a difference of two steps on the SMART scale. Finally, the decision maker has to establish the weights of the new criteria. These considerations lead to the performance tableau of Table 2.9 and to the final tableau of Table 2.10. To some extent, this example illustrates that decision making is not a single isolated event but rather a process whereby the decision maker becomes increasingly aware of the relevant criteria and the interesting alternatives. It also shows that we can easily work with a mixture of quantitative and qualitative criteria.

Table 2.9 *Performance tableau for a car selection problem. Three mid-size cars under four quantitative and two qualitative criteria.*

Criterion	Grade	Renault Laguna 2.0 83 kW	Honda Accord 2.0 96 kW	Ford Mondeo 1.8 85 kW
Consumer price	9	41,000	44,000	40,000
Maximum speed	5	200	200	195
Acceleration	7	10.6	10.2	11.1
Cargo volume	6	452	405	480
Reliability	9	4	6	4
Ambiance	8	7	7	9

Table 2.10 *Final tableau for a car selection problem. Three mid-size cars under four quantitative and two qualitative criteria.*

Criterion	Weight	Renault Laguna 2.0 83 kW	Honda Accord 2.0 96 kW	Ford Mondeo 1.8 85 kW
Consumer price	0.262	4.5	4.0	4.5
Maximum speed	0.066	9.5	9.5	9.5
Acceleration	0.131	6	6.5	6
Cargo volume	0.093	7	7	7.5
Reliability	0.262	4	6	4
Ambiance	0.186	7	7	9
Final scores		6.1	6.5	6.5

The calculations are very simple indeed, and they should be. The leading objective of MCDA is to structure a decision problem. First, there is the screening phase where the decision makers, the alternatives, and the criteria are selected. Next, the performance tableau is drawn up. If the information so obtained does not readily bring up a preferred alternative, a full-scale MCDA problem arises. By an appropriate choice of the ranges representing the context of the actual decision problem the decision maker can easily assign impact grades to the alternatives under the respective criteria, in order to express how well the alternatives perform within the corresponding ranges. The final grades designate a possibly preferred alternative. Although the calculations are so simple that the decision maker needs in principle a pocket calculator only, we would recommend him/her to use a computer program for MCDA. Otherwise it is too easy to make an error which totally derails the discussions during the decision process.

2.7 TWO ENVIRONMENTAL CASE STUDIES

In order to illustrate the performance of SMART in comparison with other methods we consider two case studies in the management of environmental problems. We are first concerned with a particular test problem, the choice of a proper connection with a given suburb, published by Janssen and Herwijnen (1994) in order to show the performance of several methods for decision support. The methods in question, mainly ordinal ones, date from the early eighties (Voogd, 1982; Hinloopen et al., 1983, 1990; Nijkamp et al., 1990). Next, we analyze the case study of Keeney and Nair (1977), the choice of a location for a nuclear power plant, which has also been used as a test problem by several others (see Roy and Bouyssou, 1983, 1993).

The test problem of Janssen and Herwijnen (1994). Throughout the monograph the authors concern themselves with the choice between three alternative options to improve the connections with a given suburb: the construction of a new highway, the redesign of the existing roads for an improved bus connection, and the construction of a new railway. There are two versions of the problem. In the first version, the performance of the alternatives under the prevailing criteria can be expressed in physical and monetary units so that we have quantitative criteria only. Table 2.11 summarizes the information.

Table 2.11 *Performance tableau in the test problem of Janssen and Herwijnen. First version with quantitative criteria only.*

Criterion	Unit	Highway	Road/bus	Railway
Costs	mln $	200	250	400
Reduced travel time	min/prs/day	30	15	10
Capacity	mln km/year	20	30	40
NO_x emission	ton/year	1000	750	100
Landscape lost	hectare	10	12	2

The information to establish the criterion weights has to be supplied by the decision makers. Since there are several methods in the monograph to process the information there are also several slightly varying sets of weights. The criteria have a persistent rank order, however. Maximization of the capacity and minimization of the NO_x emission are leading and equally important. They are followed by cost minimization, and thereafter one finds maximization of the reduced travel time and minimization of the landscape lost, both equally important as well. Ranges for the acceptable performance data are not given so that the context of the decision problem is unknown. We have therefore chosen the grades that might have been assigned to the criteria, and we have also made a guess at the endpoints of the ranges. The results, with the criterion weights calculated according to formula (2.11), are presented in Table 2.12.

Table 2.12 *Hypothetical grades and ranges assigned to the criteria in the test problem of Janssen and Herwijnen.*

Criterion	Grade	Weight	Lower end	Upper end
Costs	6	0.191	200	800
Reduced travel time	5	0.135	10	60
Capacity	7	0.270	10	100
NO_x emission	7	0.270	0	2000
Landscape lost	5	0.135	2	20

The desired target is at the lower end of the range if the corresponding criterion aims at the minimization of certain effects (costs, emission, landscape lost). It could be at the upper end if the corresponding criterion stipulates maximization (reduced travel time, capacity), but since the maximum reduction of travel time and the maximum capacity constitute hazy targets we start the categorization of the corresponding ranges at the lower end. We now have the information to evaluate the alternative options. The cost of the road/bus option, for instance, receives the impact grade

$$10 - {}^{2}\log\left(\frac{250-200}{800-200} \times 64\right) \approx 7.5,$$

provided that we round off to the nearest multiple of 0.5, and the reduced travel time gets the impact grade

$$4 + {}^{2}\log\left(\frac{15-10}{60-10} \times 64\right) \approx 6.5.$$

Table 2.13 summarizes the evaluation. Obviously, the railway seems to be the preferred alternative. Janssen and Herwijnen (1994) came to the same conclusion, but they found that the road/bus option ranks second instead of the highway. Since the context has not explicitly been modelled in their study we cannot readily conclude that the rank order of the alternatives is method-dependent.

Table 2.13 *Final tableau for the test problem of Janssen and Herwijnen. First version with quantitative criteria only.*

Criterion	Weight	Highway	Road/bus	Railway
Costs	0.191	10	7.5	5.5
Reduced travel time	0.135	8.5	6.5	4.0
Capacity	0.270	7.0	8.0	8.5
NO_x emission	0.270	5.0	5.5	8.5
Landscape lost	0.135	5.0	5.0	10
Final scores		7.0	6.7	7.5

The second version of the test problem has some qualitative criteria: reduced travel time and landscape lost. The performance of the alternative options is now expressed on a qualitative scale. Table 2.14 presents the information and Table 2.15 the evaluation. The results of the two versions are practically the same and the railway option is leading again.

Table 2.14 *Performance tableau in the test problem of Janssen and Herwijnen. Second version with quantitative and qualitative criteria.*

Criterion	Weight	Highway	Road/bus	Railway
Costs	0.191	200	250	400
Reduced travel time	0.135	+++	++	+
Capacity	0.270	20	30	40
NO_x emission	0.270	1000	750	100
Landscape lost	0.135	- - -	- - -	-

Table 2.15 *Final tableau for the test problem of Janssen and Herwijnen. Second version with quantitative and qualitative criteria.*

Criterion	Weight	Highway	Road/bus	Railway
Costs	0.191	10	8.0	5.5
Reduced travel time	0.135	10	9	8
Capacity	0.270	7.0	8.0	8.5
NO_x emission	0.270	5.0	5.5	8.5
Landscape lost	0.135	4	4	6
Final scores		7.0	6.9	7.5

The case study of Keeney and Nair (1977). This is a well-known case study in the literature on MCDA, as we have seen in the introduction to this section. Concerned with the choice of a location for a nuclear power plant in the North-West of the USA, the authors evaluated the expected utility of nine alternative sites on the basis of the following six criteria:

1. Health and safety, the annual number of human beings possibly affected by the power plant. This number was allowed to vary over the range between 0 and 200 ($\times 10^{-3}$). In other words, there may be at most one human being affected per five years.

2. Quantity of river salmon lost by thermal pollution. This quantity was allowed to vary over the range between 0 and 300 ($\times 10^{3}$).

3. Biological impact, measured on an arithmetic scale between 0 (smallest impact) and 8 (worst possible impact), and averaged over a number of experts. Under this criterion we already have the required judgemental statements.

4. Socio-economic impact, also measured on an arithmetic scale between 0 (smallest impact) and 7 (worst possible impact), and averaged over a number of experts. Under this criterion we again have the required judgemental statements.

5. Aesthetical impact, measured by the length of the high-voltage lines connecting the plant to the electrical network. The length was allowed to vary over the range between 0 and 50 miles.

6. Cost, the incremental cost beyond the cost of the cheapest possible location. The increment was allowed to vary over the range between US$ 0 and US$ 40 million.

Table 2.16 shows the impacts of the locations under the respective criteria. Moreover, we understand that the criteria have the following rank order in importance: biological impact < aesthetical impact < socio-economic impact < river salmon lost < health and safety < cost. We have roughly modelled the rank order by the assignment of the grades 4, 5, ..., 9. Since there is a range of acceptable performance data under each criterion (the ranges were required for the construction of the utility functions) we can immediately convert the impacts into grades between 4 (unsatisfactory performance that can be compensated) and 10 (excellent performance) via the formulas in Section 2.4. The desired targets are all at the lower endpoints. Thus, an impact of 57 on the range (0, 200), when the objective is to minimize the number of individuals affected by the power plant, yields the impact grade

$$10 - {}^{2}\log\left(\frac{57}{200} \times 64\right) \approx 6.0,$$

when we round off to the nearest multiple of 0.5. Table 2.17 exhibits the impact grades so obtained, the criterion weights computed according to formula (2.11), and the final grades of the alternative locations. Lastly, Table 2.18 shows the alternative locations in the rank order obtained by Keeney and Nair via Multi-Attribute Utility Theory (MAUT), as well as our rank order. The high degree of similarity between the rank order positions is encouraging because it shows that MCDA methods may be rather robust, despite the divergence in the underlying ideas. Nevertheless, a preferred alternative does not clearly emerge. The expected utilities and the final grades are too close.

Table 2.16 *Performance tableau in the case study of Keeney and Nair. Impacts of nine alternative locations for a nuclear power plant under six performance criteria.*

Crit.	Grade	A_1	A_2	A_3	A_4	A_5	A_6	A_7	A_8	A_9
1	8	57	40	25	48	44	23	52	11	18
2	7	6	6	6	0.83	2.55	0.75	0.45	4.3	3.65
3	4	1.5	1.5	2	3.5	4.5	4.5	2.5	2	1
4	6	2.5	2.5	2.5	3.5	3	4	3.5	4	3
5	5	1	1	7	6	12	1	0	0	0
6	9	2.0	0	1.5	1.9	12.3	17.7	4.8	10.9	11.4

Table 2.17 *Final tableau in the case study of Keeney and Nair. Impact grades, criterion weights, and final grades of the alternative locations.*

Crit.	Weight	A_1	A_2	A_3	A_4	A_5	A_6	A_7	A_8	A_9
1	0.237	6	6.5	7	6	6	7	6	8	7.5
2	0.167	9.5	9.5	9.5	10	10	10	10	10	10
3	0.059	8.5	8.5	8	6.5	5.5	5.5	7.5	8	9
4	0.118	7.5	7.5	7.5	6.5	7	6	6.5	6	3
5	0.084	9.5	9.5	7	7	6	9.5	10	10	10
6	0.335	8.5	10	8.5	8.5	5.5	5	7	6	6
Final scores		8.0	8.7	8.0	7.7	6.6	6.8	7.5	7.6	7.7

Table 2.18 *Rank order of the alternative locations calculated via MAUT (expected utilities) and SMART (final grades).*

Expected utilities, MAUT		Final scores, SMART	
A_3	0.926	A_2	8.7
A_2	0.920	A_3	8.0
A_1	0.885	A_1	8.0
A_4	0.883	A_4	7.7
A_8	0.872	A_9	7.7
A_9	0.871	A_8	7.6
A_7	0.862	A_7	7.5
A_5	0.813	A_6	6.8
A_6	0.804	A_5	6.6

The treatment of ordinal information. In the above examples we have incidentally shown how ordinal information can be processed within the framework of SMART. We assigned equidistant grades with a difference of one unit to the rank-ordered stimuli. The decision maker can eventually check whether the grades assigned to the stimuli with the highest and the lowest rank-order position have a reasonable distance. If not, the distance should be adjusted until the equidistant series of assigned grades seems to be acceptable. In the absence of any further cardinal information this is a simple and attractive procedure because it enables the decision maker to explore the situation. He/she does not have to turn to more complicated ordinal methods. Note, however, that the deletion or addition of alternatives may unexpectedly affect the rank order of the remaining alternatives.

The possible benefits of MCDA. In the above test problems and case studies it was not easy to identify a preferred alternative. The final grades are fairly close. A sensitivity analysis via a fuzzy version (the author, 1997) shows that the differences between two final grades should be fairly large before definite conclusions may be drawn. This is not a coincidence but a frequently occurring phenomenon. MCDA is mostly invoked to support the decision makers in muddy situations, when there are widely varying views in the decision making body about the formulation and the importance of the criteria, for instance, or when the evaluation of the alternatives appears to be highly controversial. The real contribution of MCDA is that it may improve the quality of the decision process (it makes the criteria operational) or the quality of the decision itself (it shows the strengths and weaknesses of the alternatives and it may bring up new alternatives as soon as the framework of the problem is enlarged). It rarely happens that MCDA accelerates the decision process although such a tangible benefit (time is money) would be a strong argument to promote the analysis.

2.8 QUALITY ASSESSMENT OF ACADEMIC RESEARCH

The assignment of grades is usual in the Quality Assessment of Academic Research in the Netherlands (Economics, 1995; Mathematics and Computer Science, 1997). The International Review Committees, with seven or eight members invited by the Royal Netherlands Academy of Sciences, carefully described the echelons (excellent, good, satisfactory, unsatisfactory, and poor) of the five-point scale under each of the four criteria: the quality, the productivity, the relevance, and the long-term viability of the research programmes or the respective groups (not the scientists individually). Indeed, there was a uniform scale for the three fields under all criteria. In order to illustrate the assignment of grades we briefly summarize how the Review Committees for Economics (1995) and for Mathematics and Computer Science (1997) interpreted the echelons on the five-point scale. They used practically the same interpretation, but the Mathematics and Computer Science Committee was somewhat more verbose.

The assessment of the scientific quality of a research programme depends both on the originality of the ideas and on the level of the scientific publications. The scale is presented in Table 2.19.

Table 2.19 *Verbal scale for the assessment of academic research under the criterion of scientific quality.*

Excellent	The programme makes important and innovative contributions to its field. It is among the world's leading programmes.
Good	The programme makes worthwhile contributions to its field and may contain elements of excellence.
Average	The programme's contributions are of interest, but they do not attract international attention.
Unsatisfactory	The programme contributes marginally to its field. It needs improvement in order to contribute significantly.
Poor	The programme's contributions to its field are insignificant. The programme should be reoriented or discontinued.

The scientific productivity of a research group depends on the amount of scientific output in the scientific literature and on the size of the group. The Economics Committee used a classification of the research output items (books, articles, dissertations, etc.) in order to judge the research programmes via the scale of Table 2.20.

Table 2.20 *Verbal scale for the assessment of academic research under the criterion of scientific productivity.*

Excellent	The programme has a high output of scientific publications of high impact. It produces a considerable number of PhD dissertations.
Good	The programme has a regular output of scientific publications with a considerable impact. It successfully communicates its results to a relevant audience, as identified by its mission statement.
Average	The level of productivity is average.
Unsatisfactory	The programme has produced some scientific output, but both the number and the quality of its publications are below standard.
Poor	The programme has hardly produced scientific output and the output is of marginal interest.

In assessing the relevance the Committees considered the scientific relevance of the questions addressed in the programme, the relevance of the programme for the field, and the societal relevance of the group's research. Since there were three separate aspects or dimensions here the Economics Committee could not describe the echelons of the scale. The Mathematics and Computer Science Committee interpreted the echelons as shown in Table 2.21.

Table 2.21 *Verbal scale for the assessment of academic research under the criterion of scientific and societal relevance.*

Excellent	Notable and influential contributions have been made to prominent fields. The group plays important roles in relevant scientific and industrial communities.
Good	Either less influential contributions have been made to prominent fields or else notable and influential contributions have been made to less prominent fields.
Average	The group has performed moderately well on not very prominent but still useful sub-fields.
Unsatisfactory	The research is of no great relevance to the actual or potential applications in its field.
Poor	The research is of no relevance for its field.

The assessment of the long-term viability combines aspects of past scientific performance with scientific plans for the future and with foreseeable developments in personnel and facilities. The scale is shown in Table 2.22.

Table 2.22 *Verbal scale for the assessment of academic research under the criterion of long-term viability.*

Excellent	The programme has a well-established, leading position in its field and has clear and scientifically promising plans for the future. Availability of highly qualified staff as well as future funding are secured.
Good	The programme's approach has been fruitful. Future plans and perspectives seem to be healthy. There are no major worries concerning the availability of competent staff or future funding.
Average	There are some hesitations about one or more of the aspects just mentioned, but there is a fair expectation that the group can maintain or obtain an adequate position in its field.
Unsatisfactory	The Committee has serious doubts about the programme's future position in its field. Radical actions are necessary to secure an adequate contribution in the future.
Poor	In its present form the programme should be discontinued for several reasons.

The verbal five-point scale has been converted into a numerical five-point scale by the following assignment of values: excellent = 5, good = 4,..., poor = 1. This is not an ordinal scale because the International Review Committees used, for instance, the category of average performance (an average is a cardinal concept). There were no provisions to encode hesitations between adjacent terms or values. Such provisions could easily have been made, either by the introduction of a nine-point scale with numerical values such as 4.5 to represent a hesitation between good and excellent, etc., or by the transition to a nine-point scale such that excellent = 10, good = 8, ..., poor = 2. The even values of the last-named scale would represent the original verbal terms, the odd values would stand for the hesitations between adjacent terms. The International Review Committees did not further specify the echelons, even in those cases where they could have done so. A quantitative concept like "high output" under the criterion of scientific productivity does not have to remain as vague as it was in the assessment reports. The Mathematics and Computer Science Committee defined the terms "excellent" and "good" under the criterion of scientific quality by referring to international standards. Thus, they used a frame that did not depend on the alternatives that happened to be present in the assessment procedure.

It is interesting to note that the International Review Committees did not weigh the criteria. They left it to the universities, the faculties, the programme leaders, and the individual scientists to draw their own conclusions from the grades in the assessment reports. It is easy to imagine, however, that our National Science Foundation takes the initiative to ask for a global assessment of some or all research programmes on the basis of the grades. The objective of the NSF may be to select a preferred programme, to classify the programmes in a small number of categories, or to rank the programmes in a descending order of global quality. Since the Review Committees introduced a uniform scale to judge the programmes one can easily derive a numerical scale to quantify the relative importance (the weight ratio) of any two criteria. One only has to follow the line of reasoning in Section 2.5. Consider, for instance, two real or imaginary alternatives A_j and A_k and two criteria C_f and C_s such that the decision maker's preference for A_j over A_k

under the first criterion C_f is roughly equal to his/her preference for A_k over A_j under the second criterion C_s. Suppose that the performance of the alternatives has been expressed in grades on the five-point scale 1, 2, .., 5 or, with some refinements, on the nine-point scale 1, 1.5, 2, 2.5, ..., 5. If the impact grades of the two alternatives are 4 units apart under each of the two criteria (the maximum distance on the scale), and if the final grades are 3.5 units apart the ratio ω of the two criteria has to satisfy the relation

$$\left\{\frac{\omega}{\omega+1}g_{fj}+\frac{1}{\omega+1}g_{sj}\right\}-\left\{\frac{\omega}{\omega+1}\left(g_{fj}-4\right)+\frac{1}{\omega+1}\left(g_{sj}+4\right)\right\}=3.5, \quad (2.12)$$

whence again $\omega = 15$. So, even if the performance of the alternatives is expressed on the uniform nine-point scale 1, 1.5, 2, ..., 5, we can again employ the scale $^1/_{16}$, $^1/_8$, $^1/_4$, $^1/_2$, 1, 2, 4, 8, 16 for the gradations of relative importance.

Direct assignment of grades. The above example shows that the direct assignment of grades is supported by a careful description, possibly illustrated by examples, of what the grades mean within the context of the actual decision problem. If such a description cannot be given, the user may resort to a procedure which is even simpler. He/she may subdivide the alternatives into three categories, the good ones within the given context, the bad ones, and an intermediate category, whereafter he/she may assign the grades 8, 9, or 10 to the good ones, the grades 2 or 3 (poor performance that cannot be compensated) or 4 or 5 (unsatisfactory but compensatable performance) to the bad ones, and the remaining grades to the alternatives in the intermediate category. The underlying idea is that the user seems to know the vague boundaries between the three categories, because it frequently happens that the above subdivision is intuitively made.

Estimating the relative importance of the criteria. If it is difficult for the decision maker to estimate the relative importance of the criteria *in abstracto*, he/she may consider some more concrete examples in order to clarify matters. Suppose, for instance, that he/she has to estimate the relative importance of quality and productivity in the assessment of academic research. He/she may compare two alternatives, A_j with the respective grades 5 and 4, and A_k with the grades 4 and 5. If he/she is indifferent between these alternatives, the two criteria have the same importance. If he/she has some preference for A_j, then quality has a somewhat higher importance than productivity, etc.

REFERENCES TO CHAPTER 2

1. Birnbaum, M.H., "Controversies in Psychological Measurement". In B. Wegener (ed.), *"Social Attitudes and Psycho-Physical Measurement"*. Hillsdale, New Jersey, 1982, 401 – 485.

2. Bruce, V., and Green, P.R., *"Visual Perception, Physiology, Psychology, and Ecology"*. Lawrence Erlbaum, Hove, UK, 1990.

3. Gennip, C.G.E. van, Hulshof, J.A.M., and Lootsma, F.A., "A Multi-Criteria Evaluation of Diseases in a Study for Public-Health Planning". *European Journal of Operational Research* 99, 236 – 240, 1997.

4. Hinloopen, E., and Nijkamp, P., "Qualitative Multiple Criteria Choice Analysis". *Quality & Quantity* 24, 37 – 56, 1990.

5. Hinloopen, E., Nijkamp, P., and Rietveld, P., "Qualitative Discrete Multiple Criteria Choice Models in Regional Planning". *Regional Science and Urban Economics* 13, 77 – 102, 1983.

6. Honert, R.C. van den, and Lootsma, F.A., "Assessing the Quality of Negotiated Proposals using the REMBRANDT System". To appear in the *European Journal of Operational Research,* 1999.

7. Janssen, R., and Herwijnen, M., *"DEFINITE, a System to Support Decisions on a FINITE Set of Alternatives"*. Kluwer, Dordrecht/Boston/London, 1994.

8. Keeney, R.L., and Nair, K., "Evaluating Potential Nuclear Power Plant Sites in the Pacific North-West using Decision Analysis". In D.E. Bell, R.L. Keeney, and H. Raiffa (eds.), *"Conflicting Objectives in Decisions"*. IIASA, Laxenburg, Austria, Chapter 14, 1977.

9. Lootsma, F.A., "Scale Sensitivity in a Multiplicative Variant of the AHP and SMART". *Journal of Multi-Criteria Decision Analysis* 2, 87 – 110, 1993.

10. Lootsma, F.A., "A Model for the Relative Importance of the Criteria in the Multiplicative AHP and SMART". *European Journal of Operational Research* 94, 467 – 476, 1996.

11. Lootsma, F.A., *"Fuzzy Logic for Planning and Decision Making"*. Kluwer Academic Publishers, Boston, 1997.

12. Marks, L.E., *"Sensory Processes, the New Psycho-Physics"*. Academic Press, New York, 1974.

13. Michon, J.A., Eijkman, E.G.J., and Klerk, L.F.W. de (eds.), *"Handboek der Psychonomie"*. Van Loghum Slaterus, Deventer, The Netherlands, 1976.

14. Mousseau, V., *"Problèmes Liés à l'Evaluation de l'Importance Relative des Critères en Aide Multicritère à la Décision"*. Thèse, LAMSADE, Université de Paris-Dauphine, 1993.

15. Nijkamp, P., Rietveld, P., and Voogd, H., *"Multicriteria Evaluation in Physical Planning"*. North-Holland, Amsterdam, 1990.

16. Quality Assessment of Research, Economics. Association of the Universities in the Netherlands, Leidseveer 35, PO Box 19270, 3501 DG Utrecht, 1995.

17. Quality Assessment of Research, Mathematics and Computer Science. Association of the Universities in the Netherlands, Leidseveer 35, PO Box 19270, 3501 DG Utrecht, 1997.

18. Roberts, F.S., *"Measurement Theory"*. Addison-Wesley, Reading, MA, 1979.

19. Rosch, E., "Principles of Categorization". In E. Rosch and B. Lloyds (eds.), *"Cognition and Categorization"*. Lawrence Erlbaum, Hillsdale, New Jersey, 1978, pp. 27 – 48.

20. Roy, B., and Bouyssou, D., "Comparison of Two Decision Aid Models Applied to a Nuclear Power Plant Siting Example". Cahier 47, LAMSADE, Université de Paris-Dauphine, 1983.

21. Roy, B., et Bouyssou, D., *"Aide Multicritère à la Décision: Méthodes et Cas"*. Economica, Collection Gestion, Paris, 1993.

22. Russo, J.E., and Schoemaker, P.J.H., *"Decision Traps"*. Simon & Schuster, New York, 1989.

23. Saaty, T.L., "*The Analytic Hierarchy Process, Planning, Priority Setting, Resource Allocation*". McGraw-Hill, New York, 1980.

24. Stevens, S.S., "On the Psycho-Physical Law". *Psychological Review* 64, 153 - 181, 1957.

25. Stevens, S.S., and Hallowell Davis, M.D., *"Hearing, its Psychology and Physiology"*. American Institute of Physics, New York, 1983.

26. Torgerson, W.S., "Distances and Ratios in Psycho-Physical Scaling". *Acta Psychologica* XIX, 201 – 205, 1961.

27. Veit, C.T., "Ratio and Subtractive Processes in Psycho-Physical Judgement". *Journal of Experimental Psychology: Genral* 107, 81 – 107, 1978.

28. Vincke, Ph., *"L'Aide Multicritère à la Décision"*. Editions de l'Université Libre de Bruxelles, Belgique, 1989.

29. Voogd, H., "Multicriteria Evaluation with Mixed Qualitative and Quantitative Data". *Environment and Planning* 8, 221 – 236, 1982.

30. Walker, W., Abrahamse, A., Bolten, J., Kahan, J.P., Riet, O. van de, Kok, M., and Braber, M. den, "A Policy Analysis of Dutch River Dike Improvements: Trading Off Safety, Cost, and Environmental Impacts". *Operations Research* 42, 823 – 836, 1994.

31. Winterfeldt, D. von, and Edwards, W., *"Decision Analysis and Behavioral Research"*. Cambridge University Press, Cambridge, UK, 1986.

32. Zwicker, E., *"Psychoakustik"*. Springer, Berlin, 1982.

CHAPTER 3

THE AHP, PAIRWISE COMPARISONS

The Analytic Hierarchy Process (AHP) of Saaty (1980) is a widely used method for MCDA, presumably because it elicitates preference information from the decision makers in a manner which they find easy to understand. The basic step is the pairwise comparison of two so-called stimuli, two alternatives under a given criterion, for instance, or two criteria. The decision maker is requested to state whether he/she is indifferent between the two stimuli or whether he/she has a weak, strict, strong, or very strong preference for one of them. The original AHP has been criticized because the algorithmic steps do not properly take into account that the method is based upon ratio information. The shortcomings can easily be avoided in the Additive and the Multiplicative AHP to be discussed in the present chapter. The Additive AHP is the SMART procedure with pairwise comparisons on the basis of difference information. The Multiplicative AHP with pairwise comparisons on the basis of ratio information is a variant of the original AHP. There is a logarithmic relationship between the two methods that we will extensively employ in the elicitation of preference information..

3.1 PAIRWISE COMPARISONS

We first consider the assessment of the alternatives under the respective criteria; the evaluation of the criteria themselves will be described in Section 3.3. In the basic pairwise-comparison step of the AHP, two alternatives A_j and A_k are presented to the decision maker whereafter he/she is requested to judge them under a particular criterion. The underlying assumptions are: (a) under the given criterion the two alternatives have subjective values V_j and V_k for the decision maker, and (b) the judgemental statement whereby he/she expresses his/her relative preference for A_j with respect to A_k provides an estimate of the ratio V_j/V_k. For reasons of simplicity we immediately illustrate the basic step via the subjective evaluation of cars, first under the price criterion and thereafter under the reliability criterion. Finally, we discuss the subjective evaluation under qualitative criteria.

Cars under the price criterion. We assume again (see also Section 2.2) that the decision maker is only prepared to consider alternative cars with prices between two anchors: a lower bound P_{min}, the price to be paid anyway for the cars which he/she seriously has in mind, and an upper bound P_{max}, the price that he/she cannot or does not really want to exceed. In order to model the relative preference for A_j with respect to A_k we categorize the prices which are in principle acceptable. We first "cover" the range (P_{min}, P_{max}) by the grid with the geometric sequence of points

$$P_\nu = P_{min} + (P_{max} - P_{min}) \times \frac{2^\nu}{64}, \quad \nu = 0, 1, \ldots, 6.$$

Just like in SMART we take P_ν to stand for the ν-th price category and the integer-valued parameter ν for its order of magnitude, which is given by

$$\nu = {}^2\log\left(\frac{P_\nu - P_{min}}{P_{max} - P_{min}} \times 64\right). \tag{3.1}$$

The cars of the category P_0 are the cheap ones within the given context. The cars of the categories P_2, P_4, and P_6 are *somewhat more*, *more*, and *much more* expensive. At the odd-numbered grid points P_1, P_3, and P_5 the decision maker hesitates between two adjacent gradations of expensiveness. Sometimes we also introduce the category P_8 of the *vastly more* expensive cars which are situated beyond the range, as well as the category P_7 if the decision maker hesitates between *much more* and *vastly more* expensiveness. The even-numbered grid points are the so-called major grid points designating the major gradations of expensiveness. They constitute a geometric sequence in the range (P_{min}, P_{max}) with progression factor 4. If we also take into account the odd-numbered grid points corresponding to hesitations, we have a geometric sequence of major and threshold gradations with progression factor 2.

Suppose that the prices of the cars A_j and A_k belong to the categories represented by

$$P_{\nu_j} \text{ and } P_{\nu_k}$$

respectively. We express the relative preference for A_j with respect to A_k by the inverse ratio of the price increments above the desired target P_{min} so that it can be written as

$$r_{jk} = \frac{P_{v_k} - P_{\min}}{P_{v_j} - P_{\min}} = 2^{v_k - v_j}. \tag{3.2}$$

A car of the category P_2 is somewhat more expensive than a car in the category P_0. In other words, there is a weak preference for the car in the category P_0. It is 4 times more desirable than a car in the category P_2. On the basis of such considerations we identify weak preference with the ratio 4:1. Similarly, definite preference is identified with the ratio 16:1, and strong preference with the ratio 64:1. If the price category around P_v is represented by the grade $g = 10 - v$, the relative preference for the car A_j with respect to A_k can be expressed by the difference of grades

$$\delta_{jk} = {}^2\log r_{jk} = {}^2\log\left(\frac{P_{v_k} - P_{\min}}{P_{v_j} - P_{\min}}\right) = v_k - v_j = g_j - g_k. \tag{3.3}$$

We now put the major gradations of the decision maker's comparative judgement on a numerical scale in two different ways. We assign scale values, either to the relative preferences themselves (Multiplicative AHP), or to the logarithms of the relative preferences (Additive AHP). The assignment is shown in Table 3.1. The reader can easily complete the assignment of values to the threshold gradations between the major ones. There are now two different ways to collect the preference information from the decision maker:

Table 3.1 *Comparative judgement under the price criterion. Relative preferences with scale values assigned to them, in real magnitudes as ratios of subjective values (Multiplicative AHP), and in logarithmic form as differences of grades (Additive AHP).*

Comparative judgement of A_j with respect to A_k	Relative preference for A_j w.r.t. A_k in words	Rel. preference r_{jk} in real magnitudes	Diff. of grades $\delta_{jk} = {}^2\log r_{jk}$
A_j much cheaper	strong pref. for A_j	64	6
A_j (definitely) cheaper	strict, definite pref. for A_j	16	4
A_j somewhat cheaper	weak pref. for A_j	4	2
A_j as expensive as A_k	indifference	1	0
A_k somewhat cheaper	weak pref. for A_k	1/4	-2
A_k (definitely) cheaper	strict, definite pref. for A_k	1/16	-4
A_k much cheaper	strong pref. for A_k	1/64	-6

(a). We can ask him/her to consider the axis corresponding to the price criterion and to specify the endpoints of the range of acceptable prices. Next, we identify the judgemental categories on the range, the corresponding orders of magnitude, and the corresponding grades. Thereafter, we can immediately express his/her relative preference r_{jk} for A_j with respect to A_k under the price criterion by

$$r_{jk} = 2^{\delta_{jk}} = 2^{g_j - g_k}. \tag{3.4}$$

This context-related scaling is feasible because the performance of the alternatives under the price criterion is expressed on a one-dimensional scale.

(b). If the decision maker is unable or unwilling to specify the endpoints of the range of acceptable prices we can ask him/her to express his/her comparative judgement directly in words, that is, to state whether he/she is indifferent between the two alternatives under the given criterion, or whether he/she has a weak, a definite, or a strong preference for one of the two. Thereafter, we set the numerical estimate r_{jk} of his/her relative preference for A_j with respect to A_k under the price criterion to the appropriate value as shown in Table 3.1.

To illustrate matters, let us suppose that a decision maker considers two alternative cars A_j and A_k of Dfl 25,000 and Dfl 30,000 respectively. Initially, he/she is not prepared to specify the endpoints of the range of acceptable prices. Nevertheless, it remains necessary to keep a somewhat holistic view on the alternatives $A_1,\ldots\ldots, A_n$. The two alternatives A_j and A_k cannot reasonably be judged in isolation from the context of the selection problem. Hence, the decision maker first partitions the set of alternatives into three categories: the cars which are "good" (really cheap) because their prices are below Dfl 22,000 (roughly the grades 8 and higher); the cars which are "bad" (really expensive) because their prices are beyond Dfl 30,000 (roughly the grades 5 and lower); and the intermediate category with prices between the two thresholds just mentioned. On the basis of this information, the lower limit of the range of acceptable prices must be slightly below Dfl 22,000 whereas the midpoint of the range is roughly situated at Dfl 30,000 (see also the Figures 2.2 and 2.4). Let us suppose that the decision maker now specifies the interval between Dfl 20,000 and Dfl 40,000 as the range of acceptable prices. Here, the two cars have the respective grades 6 and 5 so that the relative preference for A_j with respect to A_k can now be modelled by setting r_{jk} = 2 (the difference of grades is one unit). We have extensively discussed these considerations in order to sketch how the decision maker could work with the consumer price criterion. In fact, he/she has to work on every criterion. Preference intensities cannot loosely be expressed.

Cars under the reliability criterion. Let us again suppose that the decision maker only considers cars with a reliability of at least R_{min} so that he/she is restricted to the interval (R_{min}, R_{max}), with R_{max} usually set to 100 % (see also Section 2.2). We cover the given range by the grid with the geometric sequence of points

$$R_\nu = R_{max} - (R_{max} - R_{min}) \times \frac{2^\nu}{64}, \quad \nu = 0, 1, \ldots, 6.$$

The alternatives are again compared with respect to the desired target. If we take the symbols

$$R_{\nu_j} \text{ and } R_{\nu_k}$$

to denote the reliability of the alternative cars A_j and A_k respectively, then the inverse ratio

$$r_{jk} = \frac{R_{max} - R_{v_k}}{R_{max} - R_{v_j}} = 2^{v_k - v_j} \tag{3.5}$$

of the distances with respect to the desired target R_{max} represents the relative preference for A_j with respect to A_k under the reliability criterion. The qualification "somewhat more reliable" implies that the inverse ratio of the distances to the desired target is 4:1, etc. Representing the reliability category around R_v by the grade $g = 10 - v$, we can also express the relative preference for the car A_j with respect to A_k by the difference of grades

$$\delta_{jk} = {}^2\log r_{jk} = {}^2\log\left(\frac{R_{max} - R_{v_k}}{R_{max} - R_{v_j}}\right) = v_k - v_j = g_j - g_k. \tag{3.6}$$

The assignment of numerical values to the major gradations of comparative judgement is shown in Table 3.2. The elicitation of preference information can now again be carried out in two different ways because reliability is expressed on a one-dimensional scale. We can ask the decision maker to specify the endpoints of the range of acceptable reliabilities whereafter we calculate the grades to be assigned to the alternative cars under the reliability criterion. This yields the corresponding difference of grades (Additive AHP) and the relative preference in its real magnitude (Multiplicative AHP). If the decision maker is unable or unwilling to specify the requested endpoints, however, we can use his/her comparative judgement directly. Thus, somewhat more reliability yields the relative preference 4:1 and the difference of grades 2, more reliability yields the relative preference 16:1 and the difference of grades 4, etc.

Table 3.2 *Comparative judgement under the reliability criterion. Relative preferences with scale values assigned to them, in real magnitudes as ratios of subjective values (Multiplicative AHP), and in logarithmic form as differences of grades (Additive AHP).*

Comparative judgement of A_j with respect to A_k	Relative preference for A_j w.r.t. A_k in words	Rel. preference r_{jk} in real magnitudes	Diff. of grades $\delta_{jk} = {}^2\log r_{jk}$
A_j much more reliable	strong pref. for A_j	64	6
A_j more reliable	strict (definite) pref. for A_j	16	4
A_j somewhat more reliable	weak pref. for A_j	4	2
A_j as reliable as A_k	indifference	1	0
A_k somewhat more reliable	weak pref. for A_k	1/4	-2
A_k more reliable	strict (definite) pref. for A_k	1/16	-4
A_k much more reliable	strong pref. for A_k	1/64	-6

Cars under qualitative criteria. An appealing feature of the AHP is that it can easily support the decision makers in the assessment of the alternatives under qualitative criteria, when the performance cannot be expressed in physical or monetary units on a one-dimensional scale. Typical examples of qualitative criteria in the car selection problem are comfort, ambiance, and car body design. Despite the absence of quantitative data the decision maker can sometimes easily declare that the car A_j is somewhat more, more, or much more comfortable than A_k. It may also happen that the two cars are incomparable on the basis of their design or style. Under certain angles the decision maker prefers A_j, under other angles A_k. His/her judgement may even violate the transitivity property. He/she may prefer A_j over A_k, A_k over A_l, and A_l over A_j. This so-called cyclic judgement, an indication that in principle the alternatives cannot be ranked in a subjective order of preference, is easy to understand because human beings are dynamic observers (see also Section 2.5 and 3.7). They do not consider the alternatives from a fixed viewpoint, but they move around them in order to appreciate their properties. Their conclusions depend on their viewpoints. Finally, if the decision maker cannot express his/her relative preference in the pairwise comparison of A_j and A_k under the car body design criterion we have to accept that the corresponding r_{jk} is not available. In the next section we will show that the analysis can easily proceed, even when the judgemental statements are incomplete.

Ratio and difference information. The above procedure whereby we assign numerical values to the relative preferences themselves or to the logarithms of the relative preferences is similar to the mode of operation in acoustics where ratios of sound intensities are encoded, either in real magnitudes or logarithmically on the decibel scale (logarithmic coding). The elicitation of preferential information from the decision maker seems to proceed in two different ways. We obtain ratio information on a scale with geometric progression if we ask the decision maker to formulate his/her relative preferences (Multiplicative AHP), and we obtain difference information on an arithmetic scale if we ask the decision maker to express his/her judgement via a difference of grades (Additive AHP, SMART with pairwise comparisons, the MACBETH procedure of Bana e Costa and Vansnick, 1997). For the decision maker these are alternative ways of saying the same thing, however, see Torgerson (1961) and Section 2.4 of the present volume.

3.2 CALCULATION OF IMPACT GRADES AND SCORES

In a method of pairwise comparisons we seem to collect much more information than we need. With n alternatives the decision maker is in principle invited to carry out $n(n - 1)/2$ basic experiments in order to fill the upper or the lower triangle in the matrix $\{r_{jk}\}$ of pairwise comparisons, whereas $(n - 1)$ properly chosen experiments would be sufficient (A_1 versus A_2, A_2 versus A_3, etc.). The redundancy is usually beneficial, however, since it enables us to smooth the results of the analysis. The Additive and the Multiplicative AHP can easily analyze incomplete pairwise comparisons which occur when the decision maker does not carry out the maximum number of basic experiments. In addition, they can easily be used in groups of decision makers who individually do not even carry out the minimum number of experiments.

Incomplete pairwise comparisons in a group of decision makers. Let us first consider a group of decision makers who are requested to assess the alternatives A_j and A_k under a particular criterion. We shall be assuming that these alternatives have the same subjective values V_j and V_k for all decision makers. Moreover, the decision makers are supposed to estimate the ratio V_j/V_k via their judgemental statements. These are strong assumptions, but they are not unreasonable. Many decisions are made within an organizational framework where the members have common values. Otherwise, they would not belong to the organization. This issue is discussed at length by Mintzberg (1983).

The verbal comparative judgement given by decision maker d is converted into the numerical value r_{jkd} according to the rules of the previous section, and

$$\delta_{jkd} = {}^2\log r_{jkd}.$$

Next, we approximate the vector V of subjective values via logarithmic regression. We introduce the set D_{jk} to denote the set of decision makers who actually expressed their opinion about the two alternatives under consideration, and we approximate the vector V of subjective values via the unconstrained minimization of the sum of squares

$$\sum_{j>k}\sum_{d\in D_{jk}}\left({}^2\log r_{jkd} - {}^2\log \upsilon_j + {}^2\log \upsilon_k\right)^2. \tag{3.7}$$

With the new variables

$$w_j = {}^2\log \upsilon_j$$

we can rewrite (3.7) as

$$\sum_{j>k}\sum_{d\in D_{jk}}\left(\delta_{jkd} - w_j + w_k\right)^2. \tag{3.8}$$

Obviously, it does not matter which type of information we collect from the decision makers. We minimize the sum of squares (3.8) regardless of whether we have ratio or difference information. Since (3.8) is a convex quadratic function we can easily find an optimal solution by solving the associated set of normal equations which is obtained by setting the first-order derivatives of (3.8) to zero. Using the properties

$$\delta_{jkd} = -\ \delta_{kjd} \text{ for any } j \text{ and } k,$$

and differentiating (3.8) with respect to w_j we find

$$\sum_{k=1}^{j-1}\sum_{d\in D_{jk}}\left(\delta_{jkd} - w_j + w_k\right) - \sum_{k=j+1}^{n}\sum_{d\in D_{kj}}\left(\delta_{kjd} - w_k + w_j\right) = \sum_{\substack{k=1\\k\neq j}}^{n}\sum_{d\in D_{jk}}\left(\delta_{jkd} - w_j + w_k\right) = 0,$$

so that the normal equations themselves take the form

$$w_j \sum_{\substack{k=1 \\ k \neq j}}^{n} N_{jk} - \sum_{\substack{k=1 \\ k \neq j}}^{n} N_{jk} w_k = \sum_{\substack{k=1 \\ k \neq j}}^{n} \sum_{d \in D_{jk}} \delta_{jkd}, \ j = 1, \ldots, n, \tag{3.9}$$

where N_{jk} denotes the cardinality of the set D_{jk}. The normal equations are dependent. They sum up to the zero equation (see the below example). There is at least one additive degree of freedom in the unconstrained minima of the function (3.8) because there are only differences of variables in the sum of squares. Hence, there is at least one multiplicative degree of freedom in the unconstrained minima of the function (3.7). In other words, we can only draw conclusions from differences w_j - w_k and from ratios υ_j/υ_k. Note that the decision makers have to judge more pairs if there are two or more degrees of freedom.

Let us illustrate the foregoing results with the pairwise comparisons of the Renault Laguna 2.0, the Honda Accord 2.0, and the Ford Mondeo 1.8 under the criterion of comfort. There are four decision makers, the parents and the two adult children of a family. They do not necessarily carry out all possible pairwise comparisons. Table 3.3 shows their verbal judgemental statements in logarithmic form: equally comfortable 0, somewhat more comfortable ± 2, more comfortable ± 4, much more comfortable ± 6. Obviously, the cells of the pairwise-comparison tableau are not completely filled with four entries each, and there is no information in the cells on the main diagonal.

Table 3.3 *Pairwise-comparison tableau which shows the comparative judgement of cars under the criterion of comfort in logarithmic form. There are four decision makers who do not judge all possible pairs. The solution to the associated normal equations yields SMART impact grades between* 4 *and* 10 *with an additive degree of freedom and the AHP impact scores summing up to* 1 *with a multiplicative degree of freedom.*

	Renault Laguna	Honda Accord	Ford Mondeo	solution of normal equ.	SMART grades	AHP scores
Renault	empty	-1, -3	-2, +2, -1	0	6.0	0.203
Honda	+1, +3	empty	-3	0.72727	6.7	0.336
Ford	+2, -2, +1	+3	empty	1.18182	7.2	0.461

The normal equations corresponding to the pairwise comparisons in Table 3.3 can be written in the explicit form

$$\begin{aligned} 5\,w_1 \quad - 2\,w_2 \quad - 3\,w_3 &= -5 \\ -2\,w_1 \quad + 3\,w_2 \quad - \quad w_3 &= +1 \\ -3\,w_1 \quad - \quad w_2 \quad + 4\,w_3 &= +4 \end{aligned}$$

The first equation originates from the first row in the pairwise-comparison tableau; the coefficient 5 stands for the total number of entries in the first row, the coefficients 2 and 3 for the number of elements in the second and the third cell, whereas the right-hand side element -5 represents the sum of the entries. The remaining equations are built up in a similar way. Since the equations are dependent (they sum to the zero equation) we drop one of them, and since the solutions have an additive degree of freedom we can arbitrarily choose one of the variables. Thus, setting $w_1 = 0$ and dropping the first equation we obtain the solution exhibited in Table 3.3. We use the additive degree of freedom, designated by the symbol θ, to shift the solution so that we obtain SMART impact grades

$$g_j = w_j + \theta, j = 1,\ldots, n, \tag{3.10}$$

which are nicely but somewhat arbitrarily situated between 4 and 10. Next, we compute

$$\upsilon_j = 2^{w_j}, \; j = 1,\ldots, n, \tag{3.11}$$

and we normalize these values in order to obtain AHP impact scores s_j by setting

$$s_j = \beta \upsilon_j, j = 1,\ldots, n, \tag{3.12}$$

where β stands for the normalization factor which guarantees that the impact scores sum up to 1. Hence,

$$s_j = \frac{2^{w_j}}{\sum_{j=1}^{n} 2^{w_j}}. \tag{3.13}$$

Note that both the shift and the normalization are cosmetic operations carried out in order to present the results in a more or less attractive way. It will be clear from Table 5.3 that the AHP impact scores (Multiplicative AHP) suggest more distinction between the alternatives than the SMART impact grades (Additive AHP). They are connected by the logarithmic relationship

$$\frac{s_j}{s_k} = 2^{g_j - g_k}. \tag{3.14}$$

This relationship depends neither on the choice of the shift constant θ nor on the choice of the normalization factor β. Throughout this chapter we will see that the really interesting, uniquely determined information consists of ratios of scores and differences of grades.

Complete pairwise comparisons by a single decision maker. The above general result can be simplified in special cases. Consider one single decision maker who expressed his opinion about all possible pairs of alternatives under the given criterion so that

$$N_{jk} = 1 \text{ for all } j \neq k.$$

The normal equations (3.9) take the simple form

$$(n-1)w_j - \sum_{\substack{k=1 \\ k \neq j}}^{n} w_k = \sum_{\substack{k=1 \\ k \neq j}}^{n} \delta_{jk},$$

or, equivalently,

$$nw_j - \sum_{k=1}^{n} w_k = \sum_{k=1}^{n} \delta_{jk},$$

if we take $\delta_{jj} = 0$ for all j. We use the additive degree of freedom in the solutions of this set of equations to set the sum of the variables to zero, which yields

$$w_j = \frac{1}{n} \sum_{k=1}^{n} \delta_{jk}. \tag{3.15}$$

This means that w_j is the arithmetic mean of the j-th row in the matrix of pairwise comparisons in logarithmic form, at least under the tacit assumption that the elements on the main diagonal are set to 0 so that we have indeed n elements in each row. By a proper shift of the w_j we can generate impact grades g_j which are situated between 4 and 10.

The AHP impact scores can also directly be computed when the pairwise comparisons are recorded as ratio estimates r_{jk}. By (3.11) and (3.15) it must be true that

$$\upsilon_j = \sqrt[n]{\prod_{k=1}^{n} r_{jk}}, \tag{3.16}$$

at least under the assumption that the main diagonal elements are available. We set $r_{jj} = 1$ for any j. So, υ_j is the geometric mean of the j-th row in the matrix of pairwise comparisons in real magnitudes. The AHP impact scores s_j are obtained by normalization, again a cosmetic operation to make sure that the scores add up to 1. An illustrative example is given by the complete set of pairwise comparisons of three cars under the criterion of comfort. There is one single decision maker, and the logarithms of his/her verbal judgement are shown in Table 3.4.

Complete pairwise comparisons in a group of decision makers. When all decision makers in a group of size G assess all possible pairs of alternatives under the given criterion, we can simplify formula (5.9) because $N_{jk} = G$ for all $j \neq k$. The normal equations are now given by

$$(n-1)Gw_j - G\sum_{\substack{k=1 \\ k \neq j}}^{n} w_k = \sum_{\substack{k=1 \\ k \neq j}}^{n} \sum_{d=1}^{G} \delta_{jkd},\ j = 1, \ldots, n,$$

Table 3.4 *Pairwise comparisons tableau which shows the comparative judgement of cars under the criterion of comfort in logarithmic form. There is one single decision maker who judges all possible pairs. The solution to the associated normal equations yields the SMART impact grades between* 4 *and* 10 *with an additive degree of freedom and the AHP impact scores summing up to* 1 *with a multiplicative degree of freedom.*

	Renault Laguna	Honda Accord	Ford Mondeo	arithmetic row means	SMART grades	AHP scores
Renault	0	-3	-1	-1.33333	4.7	0.091
Honda	+3	0	+2	+1.66667	7.7	0.727
Ford	+1	-2	0	-0.33333	5.7	0.182

so that they can be rewritten as

$$nGw_j - G\sum_{k=1}^{n} w_k = \sum_{k=1}^{n}\sum_{d=1}^{G} \delta_{jkd}, \quad j = 1, \ldots, n,$$

if we take $\delta_{jjd} = 0$ for all j. We use again the additive degree of freedom in the solutions to set the sum of the variables to zero, whence

$$w_j = \frac{1}{nG}\sum_{k=1}^{n}\sum_{d=1}^{G} \delta_{jkd}. \tag{3.17}$$

The w_j can equivalently be calculated in two different ways:

(a) We first replace all entries in a cell by their arithmetic mean, so that we have a group opinion about each pair of alternatives. Thereafter we calculate the arithmetic row means of the matrix of group opinions. It is tacitly assumed that there are zeroes in the cells on the main diagonal.

(b) We first calculate the arithmetic row means in the pairwise comparison matrices of the individual group members separately, with zeroes on the main diagonals, so that we obtain the impact grades assigned to the alternatives by each group member. Thereafter we calculate the arithmetic means of the impact grades.

The AHP impact scores under the given criterion can be found in a similar way. On the basis of (3.17) a solution of the logarithmic regression problem (3.8) is given by

$$\upsilon_j = \sqrt[n]{\sqrt[G]{\prod_{k=1}^{n}\prod_{d=1}^{G} r_{jkd}}}, \tag{3.18}$$

a formula which shows that the υ_j can be obtained by geometric-mean calculations. The results do not depend on the order of the operations. We can first calculate the group opinions about each pair of alternatives and the scores thereafter, and vice versa, under the assumption that there are ones on the main diagonals of all matrices (see also Ramanathan (1997) who employed goal programming to obtain the same results). The SMART impact grades g_j and the AHP impact scores s_j can finally be obtained by a proper shift of the w_j and a proper normalization of the υ_j respectively.

Power games in groups of decision makers. The results of this section have been generalized by Barzilai and Lootsma (1997) so that they can be used in groups of decision makers who have widely varying power positions. The crucial step is the assignment of power coefficients p_d to the respective group members. These coefficients, normalized so that they sum up to 1, stand for the relative power of the decision makers, the relative size of the constituency or the country represented by them, for instance. Impact grades and scores of the alternatives are obtained by the unconstrained minimization of

$$\sum_{j>k} \sum_{d \in D_{jk}} \left(\delta_{jkd} - w_j + w_k\right)^2 p_d, \tag{3.19}$$

clearly a generalization of (3.8) in the sense that each term in the sum of squares is weighted with the relative power of the decision maker who expressed the corresponding comparative judgement. When the pairwise comparisons are complete, a solution is given by

$$w_j = \frac{1}{n} \sum_{k=1}^{n} \sum_{d=1}^{G} \delta_{jkd} p_d. \tag{3.20}$$

A discussion of power relations in groups, and particularly the choice of the power coefficients, will be presented in Chapter 6.

3.3 CRITERION WEIGHTS AND AGGREGATION

Let us first introduce some notation here. There are m criteria $C_1, \ldots, C_m$. Suppose that we have obtained the SMART impact grades g_{ij} and the AHP impact scores s_{ij} of the respective alternatives A_j under criterion C_i via the solution $(w_{i1}, \ldots, w_{in})$ of a set of normal equations. Thus,

$$g_{ij} = w_{ij} + \theta_i, j = 1, \ldots, n,$$

$$s_{ij} = \beta_i \upsilon_{ij} = \beta_i 2^{w_{ij}}, \; j = 1, \ldots, n,$$

where θ_i and β_i stand for the shift constant and the normalization factor under the i-th criterion. Differences of impact grades and ratios of impact scores are connected by the logarithmic relationship

$$\frac{s_{ij}}{s_{ik}} = 2^{g_{ij}-g_{ik}}. \tag{3.21}$$

We also assume that there are criterion weights c_i expressing the relative importance of the respective criteria. They may have been obtained via the direct-rating procedure of Section 2.5, but they may also be generated via the method of pairwise comparisons to be described somewhat later in the present section.

Aggregation via arithmetic and geometric means. On the basis of (2.8) and (3.21) we can easily derive

$$\prod_{i=1}^{m}\left(\frac{s_{ij}}{s_{ik}}\right)^{c_i} = 2^{\Delta_{jk}}, \tag{3.22}$$

$$\Delta_{jk} = f_j - f_k = \sum_{i=1}^{m} c_i g_{ij} - \sum_{i=1}^{m} c_i g_{ik}. \tag{3.23}$$

The symbols f_j and f_k, clearly obtained via a so-called arithmetic-mean aggregation rule, stand for the final SMART grades. They are not unique. There is an additive degree of freedom θ_i under each criterion as well as a general degree of freedom η so that the final grade f_j is generally given by

$$f_j = \eta + \sum_{i=1}^{m} c_i\left(w_{ij} + \theta_i\right), \tag{3.24}$$

with arbitrary shift constants η, $\theta_1, \ldots, \theta_m$. The formulas (3.22) and (3.23) suggest that the difference of the final grades is the logarithm of a ratio of terminal scores according to the Multiplicative AHP. We therefore take

$$t_j = \alpha \prod_{i=1}^{m} s_{ij}^{c_i} = \alpha \prod_{i=1}^{m}\left(\beta_i \upsilon_{ij}\right)^{c_i} \tag{3.25}$$

to represent the terminal AHP score of A_j. The multiplicative factor α is used for cosmetic purposes only, to make sure that the final scores sum up to 1. The β_i stand for arbitrary normalization factors under the respective criteria. Obviously, the terminal AHP scores are calculated via a geometric-mean aggregation rule and, moreover,

$$\frac{t_j}{t_k} = 2^{f_j - f_k}, \tag{3.26}$$

regardless of the choice of the shift constants and the normalization factors. Hence, even the final SMART grades and the terminal AHP scores satisfy the logarithmic relationship that we found for the impact grades and the impact scores of the alternatives under each of the criteria separately (see (3.14) and (3.21)).

Aggregation of complete pairwise comparisons. When the pairwise comparisons are complete, in the sense that each decison maker judged every pair of alternatives under each criterion, we can simplify the calculations considerably. We take the symbol δ_{ijkd} to stand for the difference of grades assigned to the alternatives A_j and A_k under criterion C_i by decision maker d, and r_{ijkd} for the corresponding ratio of subjective values. It follows easily from (3.17) and (3.18) that the final SMART grades and the terminal AHP scores, if we ignore the shift constants and the normalization factors, can be written as

$$f_j = \frac{1}{nG}\sum_{i=1}^{m}\sum_{k=1}^{n}\sum_{d=1}^{G} c_i \delta_{ijkd}, \tag{3.27}$$

$$t_j = \sqrt[n]{\sqrt[G]{\prod_{i=1}^{m}\prod_{k=1}^{n}\prod_{d=1}^{G} r_{ijkd}^{\,c_i}}}\,. \tag{3.28}$$

Computationally, this implies that we can operate in any order without affecting the final results of the analysis. We can aggregate, first, over the decision makers so that we obtain the group opinion about every pair of alternatives under each criterion, thereafter over the criteria so that we obtain an aggregate pairwise-comparison matrix, and finally over the row means of the aggregate pairwise-comparison matrix to obtain the final grades and the terminal scores. We can also change the order of the operations, however, in order to check the correctness of the computational results. This will be demonstrated in the examples of Section 3.4 and 3.5.

Interpretation of a ratio of criterion weights. Shift constants and normalization factors do not affect ratios of criterion weights. We show this for the normalization factors only. In doing so we also find an interpretation for ratios of criterion weights in terms of substitution rates, as one might expect since these ratios are traditionally linked with the trade-offs between gains and losses during a move along indifference curves. On the basis of formula (3.25) we can generally write the terminal AHP score of an alternative A as a function of the approximations to the subjective values of A under the respective criteria. Thus, the terminal AHP score is given by

$$t(A) = t(\upsilon_1, \ldots, \upsilon_m) = \alpha \prod_{i=1}^{m} (\beta_i \upsilon_i)^{c_i}. \tag{3.29}$$

The arbitrary multiplicative factors $\alpha, \beta_1, \ldots, \beta_m$ appear in (3.29) for normalization purposes but we can equivalently say that they appear because we did not specify the units of performance measurement. The first-order partial derivatives of t take the form

$$\frac{\partial t}{\partial \upsilon_i} = \frac{c_i}{\upsilon_i} \times t,$$

whence

$$\frac{1}{\upsilon_{i_1}} \frac{\partial t}{\partial \upsilon_{i_2}} \bigg/ \frac{1}{\upsilon_{i_2}} \frac{\partial t}{\partial \upsilon_{i_1}} = \frac{c_{i_2}}{c_{i_1}} \tag{3.30}$$

for arbitrary i_1 and i_2, and regardless of the factors $\alpha, \beta_1, \ldots, \beta_m$. We can now study the behaviour of t as a function of υ_{i1} and υ_{i2} along a contour or indifference curve. In a first-order approximation, when we make just a small step, such a move proceeds in a direction which is orthogonal to the gradient of t, that is, in the direction

$$\left(\frac{\partial t}{\partial \upsilon_{i_2}}, -\frac{\partial t}{\partial \upsilon_{i_1}}\right).$$

This is shown in Figure 3.1. Traditionally (see Keeney and Raiffa, 1976) the ratio

$$\frac{\partial t}{\partial \upsilon_{i_2}} \bigg/ \frac{\partial t}{\partial \upsilon_{i_1}}$$

has been defined as the marginal trade-off or the marginal substitution rate between the two criteria under consideration. Under the geometric-mean aggregation rule, with the function t defined by (3.29), this ratio is not constant along an indifference curve. The author (1996) therefore introduced the relative substitution rate, which is based upon the observation that human beings generally perceive relative gains and losses, that is, gains and losses in relation to the levels from which the move starts. Thus, when a small step is made along an indifference curve, the relative gain (or loss) in the υ_{i1} direction and the corresponding relative loss (or gain) in the υ_{i2} direction are proportional to

$$\frac{1}{\upsilon_{i_1}} \frac{\partial t}{\partial \upsilon_{i_2}} \text{ and } \frac{1}{\upsilon_{i_2}} \frac{\partial t}{\partial \upsilon_{i_1}}$$

respectively. Under the geometric-mean aggregation rule (3.29) the substitution rate between relative gains and losses is a constant, not only along an indifference curve, but over the entire (υ_{i1}, υ_{i2}) space. It depends neither on the units of measurement nor on the values of the remaining variables υ_i, $i \neq i_1$ and i_2. Thus, we can meaningfully use the concept of the relative importance of the criteria, even in the absence of immediate context, when the alternatives are not yet available, for instance.

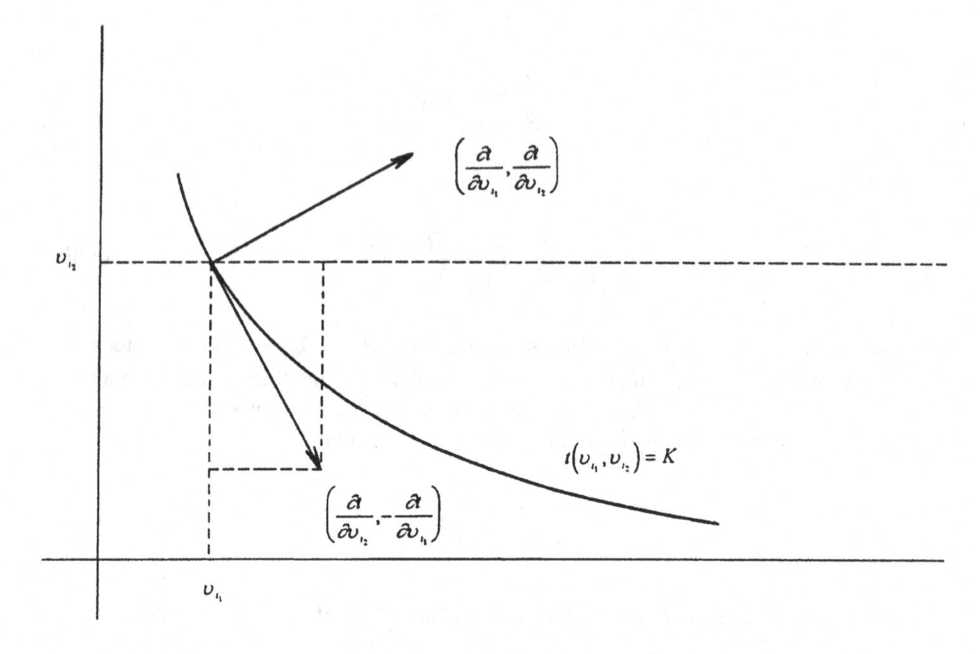

Figure 3.1 *In a first-order approximation, a move along an indifference curve proceeds in a direction which is orthogonal to the gradient in the point where the move starts. The marginal substitution rate is the ratio of the components of that direction. The relative substitution rate is the ratio of the components divided by the coordinates of the point of departure.*

Pairwise comparison of the criteria. The pairwise comparison of two criteria proceeds almost in the same way as the pairwise comparison of two alternatives under a particular criterion. It is also closely related to the procedure described in Section 2.7. In the basic experiment a pair (C_f, C_s) of criteria is presented to the decision maker whereafter he/she is requested to state whether they are equally important for him/her or whether one of the two is somewhat more, more, or much more important than the other. By the information so obtained we estimate the relative importance of the first criterion C_f with respect to the second criterion C_s, that is, the ratio of the associated criterion weights. By the arguments just mentioned the geometric-mean aggregation rule enables us to avoid the pitfalls mentioned in Section 2.5.

The assignment of numerical values to the gradations of relative importance is shown in Table 3.5. We use the scale 1, 2, 4, 8, 16 which has been derived in Section 2.5, a scale with progression factor $\sqrt{2}$ if we consider the major as well as the threshold gradations of relative importance. For reasons of simplicity we usually record the difference of grades δ_{fsd} to represent the comparative judgement of decision maker d about the pair (C_f, C_s).

Table 3.5 *Comparative judgement of the criteria. Relative importance, with scale values assigned to it in the form of a difference of grades.*

Comparative judgement of C_f with respect to C_s	Difference of grades δ_{fsd}
C_f vastly more important than C_s	8
C_f much more important than C_s	6
C_f more important than C_s	4
C_f somewhat more important than C_s	2
C_f as important as C_s	0
C_f somewhat less important than C_s	-2
C_f less important than C_s	-4
C_f much less important than C_s	-6
C_f vastly less important than C_s	-8

The criterion weights are also computed, just like the impact scores of the alternatives, via the solution of a regression problem: the unconstrained minimization of the sum of squares

$$\sum_{f>s} \sum_{d \in D_{fs}} \left(\delta_{fsd} - w_f + w_s\right)^2 . \tag{3.31}$$

The associated normal equations do not have a unique solution. There is an additive degree of freedom which can be used to obtain the normalized criterion weights

$$c_i = \left(\sqrt{2}\right)^{w_i} \Big/ \sum_{i=1}^{m} \left(\sqrt{2}\right)^{w_i} . \tag{3.32}$$

In fact, the calculation of the impact scores of the alternatives and the calculation of the the criterion weights differ with respect to the progression factor only. It is 2 for the alternatives (formula (3.13)) and √2 for the criteria (formula (3.32)), because human beings categorize long ranges on the dimensions of time, sound, and light (see Section 2.3) but evidently a short range on the dimension of the importance of the criteria.

3.4 A NOMINATION PROCEDURE

In order to illustrate the foregoing results we consider a nomination procedure for a senior professorship in Operations Research at a University in The Netherlands (the author, 1980). The decision problem in question is typically an MCDA problem. In a nomination procedure the decision makers are first concerned with the criteria. During the initial phase, when a profile is drafted and submitted to various boards, so even before there are applicants, the criteria play an important role. Finally, a choice has to be made out of three or four in principle acceptable alternatives, the candidates which are retained after the last round of interviews and discussions. The proposal to be delivered by the nomination committee consists of a short list with one or two names.

Under the Dutch legislation of the late seventies, the procedure to fill a vacant chair in a University was roughly as follows. A Nomination Committee, appointed by the Faculty Council, draws up a profile which outlines the tasks and the position of the new professor. As soon as the profile has been approved by the Faculty Council, the Board of Deans, and the Board of Management, the Committee sets out to advertise the vacancy, to invite applicants for the chair, to interview possible candidates, and to draw up a nomination proposal. The Committee has a particular authority. Even if the proposal is not acceptable for the Faculty Council, it is passed on to the Board of Deans and the Board of Management for final approval, with the critical comments of the Faculty Council and the advice given by kindred faculties enclosed. Thereafter, the proposal is submitted to the Ministry of Education for advice to the Crown. In principle, the first candidate is finally nominated by H.M. the Queen. If the first candidate withdraws his/her application, the second candidate on the list does not automatically become the preferred one. After new deliberations the Committee may come up with a new proposal, whereafter the procedure continues in the manner just decribed. Today, although some formal steps have been modified, the procedures are almost similar. Nomination committees still have a particular authority.

In the case study, the Nomination Committee consisted of seven senior professors (including the author), two associate professors, and one student, almost all taken from Operations Research Departments of Universities in our country. The first step of the analysis was the identification of the predominant criteria. According to the guidelines issued by the Ministry of Education, the committee should apply three criteria: the scientific capability of the candidates, the didactic capability, and the ability to function as a senior professor in the Faculty and the University. The Committee decided to be somewhat more specific and to consider the following criteria:

1. Didactic capability in oral communication and in the development of written material.

2. Mathematical creativity in Operations Research or in other branches of mathematics.

3. Creativity in applications of Operations Research.

4. Managerial and administrative capability, skills in project management.

5. Knowledge of Operations Research methods.

6. Experience with Operations Research in industry or in the public sector.

7. Human maturity.

These are typically qualitative criteria. In a plenary session the Committee established the relative importance of the criteria in such a cooperative manner that there was eventually complete consensus: after some vivid discussions there was exactly one group opinion about each pair of criteria. Table 3.2 shows the numerical values assigned to the verbal judgements of the Committee, not in the original form on the scale of Saaty (1980) which we will discuss in Section 3.6, but in the form of differences of grades as proposed in the previous section. Note that there are many triples of criteria which violate the cardinal consistency condition $\delta_{fj} + \delta_{js} = \delta_{fs}$. Table 3.6 also exhibits the arithmetic row means of the pairwise-comparison matrix, the grades assigned to the criteria by a shift so that the average level is 7, and the normalized criterion weights. These results (see the author, 1980) came first as a surprise, but the Committee soon agreed that the high weights of human maturity and creativity, as well as the low weights of the didactic and the administrative capabilities were not unreasonable. University professors are supposed to be loyal colleagues driven by scientific creativity. In education and administration they are amateurs. The passive knowledge of methods and applications is somewhat less important than the active creativity in new research areas. Note that practically the same criterion weights would follow from the attempt to smooth the cardinal inconsistencies by the direct assignment of grade 9 to criterion C_7, 8 to C_2 and C_3, 7 to C_6, 6 to C_1 and C_5, and 5 to C_4.

Table 3.6 *Pairwise comparison of criteria in the nomination procedure. The pairwise-comparison matrix contains estimated differences of grades. The calculated grades are obtained by a shift of the arithmetic row means, the criterion weights are derived by using the arithmetic row means as exponents of $\sqrt{2}$ because the gradations of relative importance constitute a sequence with geometric progression and progression factor $\sqrt{2}$.*

	C_1 Did. cap.	C_2 Math. creat.	C_3 Creat. appl.	C_4 Adm cap.	C_5 Knowl OR	C_6 Exp. OR	C_7 Hum mat.	arithm. row means	grades, after a shift	criterion weights, norm.
C_1	0	-2	-2	1	0	0	-4	-1.000	6.0	0.093
C_2	2	0	0	4	2	0	-2	0.857	7.9	0.177
C_3	2	0	0	1	2	2	0	1.000	8.0	0.186
C_4	-1	-4	-1	0	-1	0	-4	-1.571	5.4	0.076
C_5	0	-2	-2	1	0	-2	-2	-1.000	6.0	0.093
C_6	0	0	-2	0	2	0	-1	-0.143	6.9	0.125
C_7	4	2	0	4	2	1	0	1.857	8.9	0.250

The next step in the decision process, the evaluation of the candidates under each of the criteria separately, has not been carried out via MCDA. For the Nomination Committee the picture was sufficiently clear. The case study has been continued with three hypothetical candidates, however, in order to highlight the typical difficulties that usually arise in such a nomination procedure. The candidate A somehow represents the Operations Research practitioner with industrial experience and managerial skills. The candidate B has a long teaching experience and a broad knowledge of the existing methods in the field. The candidate C, somewhat less mature than A and B, is an active publicist in a mathematical branch of Operations Research, but he may be flexible enough to succeed in Operations

Research applications. The results of the evaluation, starting from exactly one opinion for each pair of alternatives under each criterion, are displayed in Table 3.7. Because the pairwise comparisons are complete we can calculate the final SMART grades and the terminal AHP scores in two different ways (see (3.26) and (3.27)). We can follow the usual way, that is, we first compute the arithmetic row means of the pairwise-comparison matrices, thereafter the SMART impact grades and the AHP impact scores under the respective criteria, and lastly the weighted arithmetic means of the impact grades to obtain the final results. The equivalent option is, first, to aggregate the pairwise-comparison matrices, that is, to compute weighted arithmetic means of the corresponding elements in the respective matrices, and thereafter to derive the final grades and the terminal scores directly from the arithmetic row means in the aggregate matrix. The reader may verify this in Table 3.7.

Table 3.7 *Evaluation of three candidates under seven criteria. The analysis of the aggregate matrix yields the final SMART grades and the terminal AHP scores directly.*

Criterion	weight	candidate	*A* OR pract.	*B* teaching exp.	*C* active publicist	arithm. row means	SMART grades, shifted	AHP scores, norm.
		A	0	-4	-2	-2.000	5.0	0.048
Did.cap.	0.093	*B*	4	0	2	2.000	9.0	0.762
		C	2	-2	0	0.000	7.0	0.190
		A	0	0	-1	-0.333	6.7	0.250
Math.cr.	0.177	*B*	0	0	-1	-0.333	6.7	0.250
		C	1	1	0	0.667	7.7	0.500
		A	0	2	0	0.667	7.7	0.444
Cr.appl.	0.186	*B*	-2	0	-2	-1.333	5.7	0.111
		C	0	2	0	0.667	7.7	0.444
		A	0	2	2	1.333	8.3	0.667
Adm.cap.	0.076	*B*	-2	0	0	-0.667	6.3	0.167
		C	-2	0	0	-0.667	6.3	0.167
		A	0	-1	1	0.000	7.0	0.285
Kn. OR	0.093	*B*	1	0	2	1.000	8.0	0.571
		C	-1	-2	0	-1.000	6.0	0.143
		A	0	2	2	1.333	8.3	0.667
Exp.OR	0.125	*B*	-2	0	0	-0.667	6.3	0.167
		C	-2	0	0	-0.667	6.3	0.167
		A	0	0	1	0.333	7.3	0.400
Hum. mat.	0.250	*B*	0	0	1	0.333	7.3	0.400
		C	-1	-1	0	-0.667	6.3	0.200
		A	0	0.309	0.382	0.230	7.2	0.388
Aggregate	1.000	*B*	-0.309	0	0.073	-0.079	6.9	0.314
		C	-0.382	-0.073	0	-0.151	6.8	0.298

The terminal AHP scores 0.388, 0.314, and 0.298 are so close to each other that none of the candidates is immediately eligible, but Table 3.7 provides sufficient hints and suggestions to continue the deliberations. The Committee might scrutinize the weights of the criteria C_3 and C_4, for instance, wondering whether creativity in implementations is indeed desirable for a university professor, and whether administrative skills are almost negligible. The criterion weights of Table 3.6 may reflect general priorities which are inadequate in a local situation. In a faculty where pure mathematics is predominant, for instance, attempts to corroberate the orientation towards applications may be supported, etc. A thorough review of the criterion weights might easily lead to a breakthrough in the decision process. In fact, MCDA permits decision makers to debate the reasons for their judgemental statements, to make compromises here and there, and to arrive at a consensus.

3.5 STAFF APPRAISAL

Let us also consider a somewhat related evaluation of staff members. Many industrial companies have an annual procedure to evaluate the performance and the potential of their employees. This so-called staff appraisal is sometimes based upon the comparative judgement of cohorts of employees by a panel of appraisers. The objective is to rank the employees or to classify them into a small number of categories. The information so obtained can be used to select the employees who are eligible for further management development, for an expatriate position, and/or for possible promotions. Although the companies use different names to designate the criteria for staff appraisal, they usually have similar views on the qualities which contribute to the career prospects of the employees.

In practice, the procedure is carried out via face to face interactions between appraisers and employees, and via the inspection of reports on past performance. In this section we discuss a more or less hypothetical appraisal procedure to evaluate the career potential of a cohort of the scientific staff in a research and development laboratory under the following, purely qualitative criteria:

1. Helicopter Quality: the ability to look at problems and structures from a higher viewpoint with simultaneous attention for relevant details; the ability to place facts and problems in a broader context and to choose the right course of action.

2. Conceptual Effectiveness: the ability to transform, to break down, and/or to reformulate a complicated problem into workable portions; the imaginative power, balanced by a proper sense of reality, to discern really new possibilities.

3. Operational Effectiveness: the ability of the employee to organize his/her activities, to gather the relevant knowledge and experience, to use expert advice, and to improvise when it is required; the drive to achieve substantial results.

4. Interpersonal Effectiveness (organizational effectiveness in horizontal directions): the ability of the employee to communicate his/her ideas and results, and to create a cooperative atmosphere.

5. Effective Leadership (organizational effectiveness in the vertical direction): the ability of the employee to motivate his/her subordinates and to enhance their joint performance; the drive to show a genuine respect for and interest in people.

In the first step of the appraisal procedure, for the time being in isolation from the employees to be evaluated, the relative importance of the criteria to judge the career potential in research and development was assessed by a panel of three appraisers. The results may be found in Table 3.8. Conceptual effectiveness has the highest weight, which is not surprising in a research and development laboratory. The other criteria have roughly the same weights.

Table 3.8 *Pairwise comparison of criteria in an appraisal procedure. The pairwise-comparison matrix contains estimated differences of grades.*

	C_1 Heli quality	C_2 Conc. eff.	C_3 Oper. eff.	C_4 Interp. eff.	C_5 Eff. leader	arithm. row means	grades, after a shift	criterion weights, norm.
C_1	0	-2	1	1	1	0.200	7.2	0.204
C_2	2	0	2	1	3	1.600	8.6	0.331
C_3	-1	-2	0	1	1	-0.200	6.8	0.177
C_4	-1	-1	-1	0	0	-0.600	6.4	0.154
C_5	-1	-3	-1	0	0	-1.000	6.0	0.134

There are four appraisees $A_1, \ldots, A_4$, who are subsequently judged as follows. Under the important criteria of Helicopter Quality and Conceptual Effectiveness the appraisers do not try to attain unanimity, and they sometimes abstain from expressing their comparative judgement. Table 3.9 shows their judgement under the first criterion.

Table 3.9 *Pairwise-comparison tableau which shows the comparative judgement of four appraisees under the criterion of Helicopter Quality.*

	A_1	A_2	A_3	A_4	solution of normal eq.	SMART grades	AHP scores
A_1	empty	0	+1, +1	+2, -1, 0	0.000	7.2	0.273
A_2	0	empty	+1	+2	0.524	7.8	0.393
A_3	-1, -1	-1	empty	+1, -1	-0.762	6.5	0.161
A_4	-2, +1, 0	-2	-1, +1	empty	-0.667	6.6	0.172

The system of normal equations associated with the pairwise-comparison tableau has the following explicit form (see also Section 3.2):

$$\begin{aligned}
6\,w_1 \; - \; w_2 \; - 2\,w_3 \; - 3\,w_4 &= +3 \\
-\,w_1 \; + 3\,w_2 \; - \; w_3 \; - \; w_4 &= +3 \\
-2\,w_1 \; - \; w_2 \; + 5\,w_3 \; - 2\,w_4 &= -3 \\
-3\,w_1 \; - \; w_2 \; - 2\,w_3 \; + 6\,w_4 &= -3
\end{aligned}$$

A solution is easily obtained. Since the system is dependent, we can arbitrarily set one of the variables to zero, and we can drop one of the equations. Thereafter, we arbitrarily shift the solution so that the grades have the average level 7. The Tables 3.10 - 3.13 exhibit the comparative evaluations of the appraisees under the remaining criteria. Note that the appraisers were in full agreement under the criteria of Operational Effectiveness, Interpersonal Effectiveness, and Effective Leadership.

Table 3.10 *Pairwise-comparison tableau which shows the comparative judgement of four appraisees under the criterion of Conceptual Effectiveness.*

	A_1	A_2	A_3	A_4	solution of normal eq.	SMART grades	AHP scores
A_1	empty	-1	+2, +4, +4	+1, +3	0.000	8.1	0.345
A_2	+1	empty	+4	+2	0.609	8.7	0.527
A_3	-2, -4, -4	-4	empty	-1	-3.261	4.9	0.036
A_4	-1, -3	-2	+1	empty	-1.913	6.2	0.092

Table 3.11 *Pairwise-comparison matrix which shows the comparative judgement of four appraisees under the criterion of Operational Effectiveness.*

	A_1	A_2	A_3	A_4	arithmetic row means	SMART grades	AHP scores
A_1	0	0	0	-3	-0.75	6.3	0.091
A_2	0	0	0	-3	-0.75	6.3	0.091
A_3	0	0	0	-3	-0.75	6.3	0.091
A_4	+3	+3	+3	0	2.25	9.3	0.727

Table 3.12 *Pairwise-comparison matrix which shows the comparative judgement of four appraisees under the criterion of Interpersonal Effectiveness.*

	A_1	A_2	A_3	A_4	arithmetic row means	SMART grades	AHP scores
A_1	0	+1	-1	0	0	7.0	0.222
A_2	-1	0	-2	-1	-1	6.0	0.111
A_3	+1	+2	0	+1	+1	8.0	0.444
A_4	0	+1	-1	0	0	7.0	0.222

Table 3.13 *Pairwise-comparison matrix which shows the comparative judgement of four appraisees under the criterion of Effective Leadership.*

	A_1	A_2	A_3	A_4	arithmetic row means	SMART grades	AHP scores
A_1	0	-5	-3	-3	-2.75	4.3	0.023
A_2	+5	0	+1	+1	1.75	8.8	0.530
A_3	+3	-1	0	0	0.50	7.5	0.223
A_4	+3	-1	0	0	0.50	7.5	0.223

The final grades (again with the average level 7) and the terminal scores are shown in Table 3.14. The rank order $A_2 > A_4 > A_1 > A_3$ which follows from the procedure may not have been established without unquestionable doubt because the grades and scores are fairly close, but the procedure itself may have triggered the right questions among the appraisers (do we agree that A_4 ranks second, for instance, mainly because of his/her high score for operational effectiveness?). Moreover, the procedure enables the appraisers to keep a more holistic view on the evaluation process, even if they sometimes concentrate on a detailed comparison of the alternatives.

Table 3.14 *Final tableau of the appraisees, with SMART final grades and AHP terminal scores to express overall performance under the five criteria simultaneously.*

Criterion	Weight	A_1	A_2	A_3	A_4
Helicopter Quality	0.204	7.226	7.750	6.464	6.560
Conceptual Effectiveness	0.331	8.141	8.750	4.881	6.228
Operational Effectiveness	0.177	6.250	6.250	6.250	9.250
Interpersonal Effectiveness	0.154	7.000	6.000	8.000	7.000
Effective Leadership	0.134	4.250	8.750	7.500	7.500
SMART final grades		6.923	7.680	6.278	7.119
AHP terminal scores		0.223	0.378	0.143	0.256

3.6 THE ORIGINAL AHP

Pairwise comparisons have been known for many decades. The original idea is due to Thurstone (1927). The method of Bradley and Terry (1952), see also David (1963), was not so easy to implement, and it could only be used in groups because it did not elicitate the gradations of preferential judgement. The original AHP of Saaty (1977, 1980) has been very popular from the beginning, but it has also been criticized for various reasons: (a) the so-called fundamental scale to quantify verbal comparative judgement produces inconsistencies which are not necessarily present in the decision maker's mind, (b) it estimates the impact scores of the alternatives by the Perron-Frobenius eigenvector of the pairwise-comparison matrix but it does not specify whether the left or the right eigenvector should be used, and (c) it calculates the terminal scores of the alternatives via the arithmetic-mean aggregation rule so that the results depend on the order of the computations. The original AHP is based upon ratio information so that the proposed algorithmic operations are inappropriate. Let us briefly reconsider the critical comments and the supporting evidence.

The fundamental scale. For a decision maker there is no difference between the original AHP and the Additive or the Multiplicative AHP as far as the input of the judgemental statements is concerned. Two stimuli are presented to him/her, whereafter he/she is invited to answer the same questions regardless of the method to be used. The difference is in the subsequent quantification of the answers. Table 3.15 shows how the major gradations of the decision maker's comparative judgement are encoded.

Table 3.15 *Numerical scales to quantify comparative preferential judgement in the original AHP and in two variants: the Additive and the Multiplicative AHP.*

Comparative preferential judgement of A_j with respect to A_k	Original AHP, estimated ratio of subjective values	Additive AHP, difference of grades	Multiplicative AHP, estimated ratio of subjective values
Very strong (absolute) preference for A_j	9	8	256
Strong preference for A_j	7	6	64
Strict (definite) preference for A_j	5	4	16
Weak preference for A_j	3	2	4
Indifference between A_j and A_k	1	0	1
Weak preference for A_k	1/3	-2	1/4
Strict (definite) preference for A_k	1/5	-4	1/16
Strong preference for A_k	1/7	-6	1/64
Very strong (absolute) preference for A_k	1/9	-8	1/256

The scale of the Multiplicative AHP, based upon psycho-physical arguments (Chapter 2), is clearly much longer than the fundamental scale of the original AHP. Several numerical studies, however, show that the terminal scores are almost insensitive to the length of the scale (the author, 1993; Triantaphyllou et al., 1994; Budescu et al., 1996). The fundamental scale, neither arithmetic nor geometric, introduces frictions which are not necessarily present in the decision maker's mind. Consider three alternatives A_j, A_k, and A_l, for instance, and suppose that the decision maker has a weak preference for A_j with respect to A_k (fundamental-scale value 3) and a weak preference for A_k with respect to A_l (the same fundamental-scale value 3). He/she might logically have a strict preference for A_j with respect to A_l (transitivity of preference, weak $\times$ weak $\cong$ strict), certainly not the very strong or absolute preference which is suggested by the product of the scale values for weak preference ($3 \times 3 = 9$). Frictions to such an extent do not occur under the Additive AHP ($2 + 2 = 4$, addition of logarithms) or the Multiplicative AHP ($4 \times 4 = 16$).

The Perron-Frobenius eigenvector. It is well-known that Saaty (1977, 1980), considering the positive and reciprocal matrix R with complete pairwise comparisons under a given criterion (one single decision maker), proposed to estimate the impact scores of the alternatives by the components of the eigenvector corresponding to the largest eigenvalue (real and positive by the theorem of Perron and Frobenius). At an early stage, this proposal has been criticized by Johnson et al. (1979), Cogger and Yu (1983), Crawford and Williams (1985), and Takeda et al. (1987). The key issue is the so-called right-left asymmetry: which eigenvector should be used to produce the impact scores of the alternatives under a given criterion, the left or the right eigenvector? Johnson et al. (1979) considered the pairwise-comparison matrix

$$R = \begin{pmatrix} 1 & 3 & 1/3 & 1/2 \\ 1/3 & 1 & 1/6 & 2 \\ 3 & 6 & 1 & 1 \\ 2 & 1/2 & 1 & 1 \end{pmatrix}.$$

The Perron-Frobenius right eigenvector, positive and normalized in the sense that the components sum up to 1, is given by (0.184, 0.152, 0.436, 0.227) so that it provides the rank order $A_3 > A_4 > A_1 > A_2$. The element r_{jk} in R tells us how strongly the decision maker prefers alternative A_j over A_k, and the components of the right eigenvector stand for the "degree of satisfaction" with the alternatives. If we rephrase the questions submitted to the decision maker, that is, if we ask how strongly he/she dislikes A_j with respect to A_k, we would logically obtain the transpose of R (if A_j is 3 times better than A_k, for instance, it is 1/3 times worse, etc.). The right eigenvector of the transpose of R is the left eigenvector of R itself, and it appears to be given by (0.248, 0.338, 0.105, 0.259). The components represent the "degree of dissatisfaction" with the alternatives. This leads to the rank order $A_3 > A_1 > A_4 > A_2$, and accordingly to an interchange of the positions of A_1 and A_4. Note that such a rank reversal does not occur when one uses the geometric row means to compute the impact scores. In a reciprocal matrix the geometric column means are the inverses of the corresponding geometric row means.

Saaty and Vargas (1984) tried to settle the matter and to show that the right eigenvector should always be used because it properly captures relative dominance. The evidence is not convincing, however. The basic step in the argumentation is that the (j, k)-th element in the m-th power of a pairwise-comparison matrix represents the total intensity of all walks of length m from a node j to a node k. The intensity of an m-walk from j to k is the product of the intensities of the arcs in the walk. In terms of the AHP, the total intensity would be a sum of products of preference ratios. This is a peculiar quantity, however, without a proper physical or psychological interpretation.

Aggregation of ratio and difference information. In recent years Barzilai et al. (1987, 1990, 1994, 1997) analyzed the AHP under the requirement that the results of the aggregation step should not depend on the order of the computations. The terminal scores should be the same, regardless of whether one combines first the pairwise-comparison matrices into an aggregate matrix, or whether one computes first the impact scores from the separate pairwise-comparison matrices. The original AHP is based upon ratio information so that it has a multiplicative structure. The geometric-mean aggregation rule appears to be the only rule here which satisfies certain consistency axioms (see formula (3.18)). For a variant which is based upon difference information so that it has an additive structure the arithmetic-mean aggregation rule appears to be the only rule which is compatible with the corresponding consistency axioms (see formula (3.17)).

The original AHP follows an inappropriate sequence of operations, particularly when it is used in group decision making: geometric-mean computations to synthesize the pairwise comparisons expressed by the individual members into group pairwise comparisons, eigenvector calculations to compute the impact scores of the alternatives, and arithmetic-mean calculations to combine the impact scores into terminal scores. We avoid the dependence on the order of the computations: in the Multiplicative AHP by using a sequence of geometric-mean calculations (formula (3. 28)), and in the Additive AHP by using a sequence of arithmetic-mean computations (formula (3. 27)).

Numerical experiments. How do the shortcomings of the original AHP emerge? How do the decision makers notice that there is an inappropriate sequence of operations in it? Not by inspection of the terminal scores because the AHP is remarkably robust: the terminal scores are almost insensitive to the choice of the algorithmic operations. Analyzing the example of the nomination procedure in Section 3.4 via the original AHP, with the fundamental scale to quantify the verbal opinions expressed in the Nomination Committee, we obtain the terminal scores (0.367, 0.319, 0.313). They practically coincide with the results (0.388, 0.314, 0.298) via the Multiplicative AHP.

The vivid discussions about the shortcomings of the original AHP were triggered by Belton and Gear (1983) who studied the behaviour of the method on an artificial numerical example. They noted that the addition of a copy of an alternative may change the rank order of the terminal scores in a set of consistently assessed alternatives, even when the criteria and the criterion weights remain the same. This prompted Dyer (1990) to state that the rankings provided by the original AHP are arbitrary. It is easy to verify, however, that the rank reversal in the example disappears as soon as the arithmetic-mean aggregation is replaced by the geometric-mean aggregation. In the Additive and the Multiplicative AHP the addition or the deletion of an alternative, whether it is a copy of another alternative or not, preserves the rank order of the remaining alternatives. These two variants have an even stronger property. The Additive AHP preserves the difference between any two final grades, the Multiplicative AHP preserves the ratio of any two terminal scores.

Let us clarify matters by considering here the example of Belton and Gear (1983). Table 3.16 shows the original data. There were initially three alternatives, A_1, A_2, and A_3, and three criteria with equal weights. Under each criterion the pairwise comparisons were consistent in the multiplicative sense that $r_{jk} \times r_{kl} = r_{jl}$ for any triple of elements in the pairwise-comparison matrix (hence the somewhat unusual entries 8/9 and 9/8). The terminal scores appear to be (0.45, 0.47, 0.08) so that A_2 turns out to be the preferred alternative. When a copy A_4 of A_2 is added to the set, however, the original AHP yields the terminal scores (0.37, 0.29, 0.06, 0.29) so that it designates A_1 to be the leading alternative. Table 3.17 shows that the computations in the Additive and the Multiplicative AHP are much simpler because we can use the aggregate matrix to find the final grades and the terminal scores. First, note that the pairwise-comparison matrices are consistent in the additive sense so that the aggregate matrix is also consistent. We therefore need the top row only. The remaining rows do not provide any additional information. Hence, the final grades of A_1 and A_2 have the difference -0.333, the final grades of A_1 and A_3 have the difference 5, etc., and the alternatives A_1, A_2, and A_3 have the terminal scores (0.436, 0.550, 0.014). When A_4 is added to the set, the terminal scores are (0.281, 0.355, 0.009, 0.355). No rank reversal! The ratio of any two terminal scores is preserved.

Table 3.16 *Example of Belton and Gear* (1983) *in its original form.*

Criterion		A_1	A_2	A_3	A_4
	A_1	1	1/9	1	1/9
C_1	A_2	9	1	9	1
	A_3	1	1/9	1	1/9
	A_4	9	1	9	1
	A_1	1	9	9	9
C_2	A_2	1/9	1	1	1
	A_3	1/9	1	1	1
	A_4	1/9	1	1	1
	A_1	1	8/9	8	8/9
C_3	A_2	9/8	1	9	1
	A_3	1/8	1/9	1	1/9
	A_4	9/8	1	9	1

Table 3.17 *Example of Belton and Gear* (1983) *with rescaled data. Additive AHP: pairwise comparisons as differences of grades.*

Criterion		A_1	A_2	A_3	A_4
	A_1	0	-8	0	-8
C_1	A_2	8	0	8	0
	A_3	0	-8	0	-8
	A_4	8	0	8	0
	A_1	0	8	8	8
C_2	A_2	-8	0	0	0
	A_3	-8	0	0	0
	A_4	-8	0	0	0
	A_1	0	-1	7	-1
C_3	A_2	1	0	8	0
	A_3	-7	-8	0	-8
	A_4	1	0	8	0
	A_1	0	-0.333	5	-0.333
Aggregate	A_2	0.333	0	5.333	0
	A_3	-5	-5.333	0	-5.333
	A_4	0.333	0	5.333	0

The aggregate matrix is a powerful instrument for the analysis via the Additive and the Multiplicative AHP, but it does not make sense in the original AHP. With the data of Table 3.16 it would not even be reciprocal so that the theorem of Perron and Frobenius does not apply. The key question is the following one, however. Given the assumption that the criteria and the criterion weights do not change, could we legitimately expect rank reversal by the addition or the deletion of copies of an alternative? We do not think so. Rank reversal cannot logically be expected under such circumstances so that the example of Belton and Gear (1983) provides the strong warning that the original AHP should be used with considerable caution (see also French, 1988; Stewart, 1992). Rank reversal undermines the confidence in MCDA that the decision makers may initially have (see the case study by the author et al., 1990).

3.7 INTRANSITIVITY, CYCLIC JUDGEMENT, HIERARCHIES

The evaluation of the alternatives under a particular criterion depends on the decision maker's perspective, that is, on the choice of a proper viewpoint and on the dimensions of the space that he/she overviews (Section 1.5 and Chapter 5). This hypothesis has several implications for the evaluation of the alternatives under a particular criterion whereby we generate the so-called impact grades or scores. The evaluation depends largely on the type of criterion: quantitative or qualitative, one-dimensional or low-dimensional?

Quantitative criteria are usually one-dimensional: the performance of the alternatives can be expressed in physical or monetary values on a one-dimensional scale. The decision maker's viewpoint is at one of the endpoints of the range of acceptable performance data. Convenient evaluation methods are direct rating via the assignment of grades and methods for pairwise comparisons. Qualitative criteria, however, are often low-dimensional: the performance of the alternatives can only be expressed in several more or less verbal terms. The pleasure of driving a car, for example, depends on the maximum speed, the maximum acceleration, and the level of noise and vibrations. The aesthetic attractiveness of the car body design depends on the decision maker's varying viewpoint. Convenient evaluation methods are direct rating, pairwise comparisons, or just ranking so that we have ordinal information only.

The Additive and the Multiplicative AHP can easily "solve" the paradox of cyclic judgement, as the evaluation of the alternatives $A_1,\ldots, A_4$ under an unspecified criterion in Table 3.18 readily shows. We leave it to the reader to generate the normal equations associated with the pairwise comparisons, to set the last variable to zero, and to drop the last equation. The resulting grades and scores introduce a complete preference order. The appearance of a cycle urges some decision makers to review their judgement and to ask for a recalculation of the grades and scores. Others, however, accept the contradictions in their judgemental statements as well as the "solution" presented by the analysis.

Table 3.18 *Pairwise-comparison tableau which shows the comparative judgement of four alternatives under an unspecified criterion. The pairwise comparisons are incomplete and they contain the preference cycle* $A_2 > A_3 > A_4 > A_2$.

	A_1	A_2	A_3	A_4	solution of norm. equ.	SMART grades	AHP scores
A_1	empty	+1	empty	+2	2.0	8.0	0.364
A_2	- 1	empty	+2	- 1	1.0	7.0	0.182
A_3	empty	- 2	empty	+3	1.0	7.0	0.182
A_4	- 2	+1	- 3	empty	0.0	6.0	0.091

MCDA is sometimes referred to as the collection of methods for multi-dimensional evaluation or multi-dimensional scaling. The evaluation of the alternatives under one of the criteria individually is not necessarily one-dimensional, as we have just seen. Under the qualitative criteria it is almost always low-dimensional. In such cases the construction of a hierarchy of criteria and sub-criteria is appealing because it breaks down a low-dimensional evaluation into a set of one-dimensional evaluations. The hierarchical representation is not necessarily unique, however. This may have far-reaching effects for the evaluation in its totality. Let us illustrate the construction and its possible pitfalls via the staff appraisal procedure in Section 3.5. The criteria are qualitative, and the description suggests that they are at least two- or three-dimensional. Helicopter Quality, for instance, consists of two sub-criteria: (a) the ability to look at problems and structures from a higher viewpoint, and (b) the ability to choose the right course of action. There are plausible arguments, however, to allocate the second sub-criterion to Conceptual Effectiveness or to Operational Effectiveness. This merely shows that the hierarchical tree of criteria and sub-criteria can be structured in several ways. In many decision problems the structure is the subject of intense debates because different structures lead to different sets of weights. In order to demonstrate this we limit ourselves to an appraisal procedure with two criteria and the following sub-criteria only:

Conceptual effectiveness:
(a) the analytical ability to break down a problem,
(b) the imaginative power to discern new possibilities,
(c) the fortunate skill to choose the right course of action.

Operational Effectiveness
(a) the ability to organize the activities,
(b) the ability to gather knowledge,
(c) the ability to use expert advice.

Let us suppose that the criteria are felt to be equally important so that they have a weight of 0.50 each. Moreover, let us assume that the sub-criteria under each of the respective criteria are equally important as well. Following the mode of operation in the original AHP we could accordingly assign a weight of $0.50 \times 0.333 = 0.167$ to each subcriterion. This is graphically illustrated in Figure 3.3. It is easy to imagine what happens as soon as we reallocate the skill to choose the right course of action. Let us disconnect it from

Conceptual Effectiveness in order to assign it to Operational Effectiveness. Under the assumption that the two criteria are still equally important and that the sub-criteria have the same importance under their respective criteria, we obtain a weight of $0.50 \times 0.50 = 0.25$ for the two remaining sub-criteria under Conceptual Effectiveness. Similarly, we find a weight of $0.50 \times 0.25 = 0.125$ for each of the four sub-criteria under Operational Effectiveness. Thus, the structure of the hierarchy affects the criterion weights, see also Borcherding and von Winterfeldt (1988), as well as Pöyhönen and Hämäläinen (1998). One may argue that the reallocation of the ability to choose the right course of action affects the relative importance of the two criteria. Conceptual Effectiveness has a smaller number of sub-criteria (dimensions) than Operational Effectiveness as soon as the shift has been carried out. Hence, assigning a weight of 0.167 to each of the sub-criteria with the argument that they still have the same importance we would obtain a weight of 0.333 for Conceptual Effectiveness and 0.667 for Operational Effectiveness. The argument is not strong, of course. In principle, it replaces the rather simple top-down evaluation of criteria and sub-criteria by a more laborious bottom-up evaluation. This all means that the introduction of the level of the sub-criteria does not necessarily simplify the evaluation of the alternatives. Hierarchies of criteria and sub-criteria seem to be very popular in practical applications but their structure is debatable and there are still several theoretical questions to be solved. In the Sections 2.5 and 3.3 we only formalized the concept of the relative importance of the criteria, via a model which is based upon the pairwise comparison of real or imaginary alternatives. The concept of the relative importance of the sub-criteria is still undefined, however. The original AHP disregards these issues. It carries out the evaluation of the criteria and the sub-criteria in the same way as the evaluation of the alternatives, without any argument to justify this mode of operation. In the Additive and the Multiplicative AHP, the evaluation of the alternatives and the evaluation of the criteria are slightly different. They do not use the same progression factor. The evaluation of the sub-criteria still has to be studied, however. For the time being we do not therefore recommend to apply the AHP beyond the two-level approach that we followed so far.

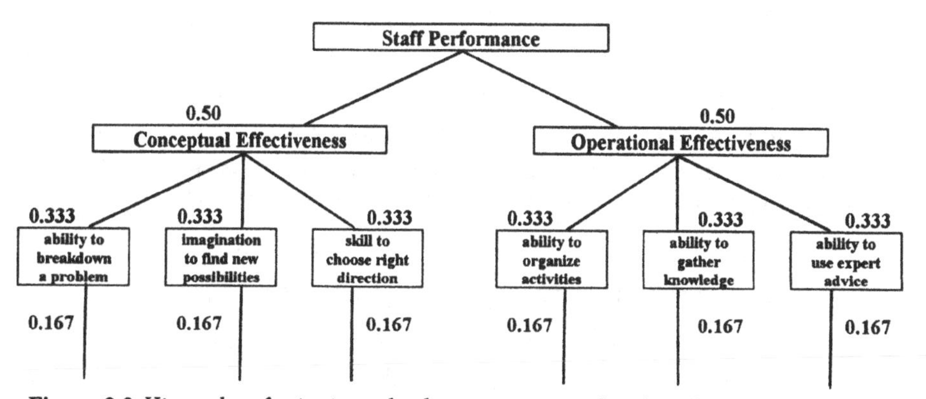

Figure 3.3 *Hierarchy of criteria and sub-criteria in a reduced staff appraisal problem. The alternatives are to be evaluated under six sub-criteria which have equal weights.*

Let us conclude this section with the "anomaly" which arises in the example of Table 3.19. The rank order $A_1 > A_2 > A_3 > A_4$ which follows from the ordinal information (preference without gradations) is in conflict with the rank order $A_1 > A_3 > A_2 > A_4$ which follows from the cardinal information (gradations or intensities of preference). This is due, either to the exaggerated preference for A_3 with respect to A_4, or to the moderate preference intensities in the other pairwise comparisons (the MCDA specialist cannot decide where errors have possibly been made). In tournaments it may also happen that the results depend on how the game is played: match play or stroke play, the rank orders are not always the same. The example shows that the identification of cycles is not enough to support the decision maker in the identification of "irrational" judgement. A simple and effective approach is to compare the original (Table 3.19) and the computed (Table 3.20) differences of grades in the upper or the lower triangle of the pairwise-comparison matrices. If there is a change of sign (see the element (2, 3)) or a large deviation (see the elements (1, 3) and (3, 4)), the decision maker may be asked to reconsider his/her judgement. We have the impression that the decision makers are mostly prepared to correct the cycles in their judgement, but they seem to be unperturbed whenever inconsistencies without cyclic judgement appear in their preference intensities.

Table 3.19 *Pairwise-comparison matrix which shows the comparative judgement of four alternatives under an unspecified criterion. The rank order $A_1 > A_2 > A_3 > A_4$ which follows from the ordinal information (preference without gradations) is in conflict with the rank order $A_1 > A_3 > A_2 > A_4$ which follows from the cardinal information (gradations or intensities of preference) in the pairwise comparisons.*

	A_1	A_2	A_3	A_4	arithmetic row means	SMART grades	AHP scores
A_1	0	+1	+2	+3	1.50	8.50	0.421
A_2	- 1	0	+1	+2	0.50	7.50	0.210
A_3	- 2	- 1	0	+8	1.25	8.25	0.354
A_4	- 3	- 2	- 8	0	-3.25	3.75	0.016

Table 3.20 *Pairwise-comparison matrix with computed differences of grades. There are significant deviations from the original differences of grades as shown in Table* 3.19: *a change of sign in cell* (2, 3) *and large deviations in the cells* (1, 3) *and* (3, 4).

	A_1	A_2	A_3	A_4	SMART grades
A_1	0	+ 1.00	+ 0.25	+ 4.75	8.50
A_2	- 1.00	0	- 0.75	+ 3.75	7.50
A_3	- 0.25	+ 0.75	0	+ 4.50	8.25
A_4	- 4.75	- 3.75	- 4.50	0	3.75

3.8 MCDA IN A DECISION TREE

Let us finally turn to applications of MCDA in situations with uncertain outcomes. A decision tree is a well-known tool to model and to evaluate a decision process which consists of an alternating sequence of actions and uncertain consequences (Keeney and Raiffa, 1976; Von Winterfeldt and Edwards, 1986). We concern ourselves here with a tree of the simplest possible form. At the initial node, the decision maker has the choice between the sure outcome A_0 on the one hand and the chance fork with the uncertain, mutually exclusive outcomes A_1 and A_2 on the other. It is customary in Decision Analysis to express the consequences in monetary values. If the value of A_0 is situated between the values of A_1 and A_2 the decision maker is supposed to estimate the probabilities p_1 and p_2 of the respective outcomes A_1 and A_2. Thereafter he/she faces the question of how to judge the chance fork, that is, how to aggregate the values of A_1 and A_2 and their associated probabilities into a single quantity that can properly be compared with a single quantity representing the sure outcome A_0. The aggregation procedure is the core of the backfolding process in larger decision trees whereby the decision maker repeatedly replaces chance forks by single quantities. During the process he/she works backwards through the tree until he/she can identify the action to be chosen at the initial node and at subsequent decision nodes.

In the aggregation procedure the decision maker first determines a range (L, H) of relevant values on the axis in question, to be used uniformly throughout the analysis. Thereafter, he/she replaces each outcome A_j $(j = 0, 1, 2)$ by a lottery or gamble with the probability u_j to obtain the high endpoint H (generally high benefits) of the range and the probability $1 - u_j$ to obtain the low endpoint L (low benefits or even losses). This is typically a laboratory experiment, of course. The decision maker chooses the so-called utility u_j in such a way that he/she is indifferent between the monetary value of A_j and the gamble. Finally, he/she compares the utility u_0 of A_0 with the subjective expected utility

$$p_1 u_1 + p_2 u_2$$

of the chance fork. One usually observes that the decision maker is risk-averting, in the sense that the monetary value of A_j is lower than the expected value

$$u_j H + (1 - u_j) L$$

of the equivalent gamble with the probability u_j. The opposite phenomenon is also known, however: if the monetary value of A_j is higher than the expected value, the decision maker is risk-seeking. The discrepancy between the monetary value of A_j and the expected value of the equivalent gamble motivated the development of Multi-Attribute Utility Theory (MAUT). There may be a particular reason for the discrepancy. Confronted with a gamble which leads either to high benefits or to low benefits or even losses, the decision maker faces an MCDA problem anyway, whether the original outcomes have one-dimensional consequences or not. He/she has to weigh, not only the benefits versus the losses with due regard to his/her financial position, but also the consequences for his/her status and prestige and for his/her relations with other people, for instance. Instead of establishing the utilities of the outcomes in a given problem the decision maker might as well concentrate on the underlying reasons why he/she is risk-averting or risk-seeking and on the analysis of the underlying MCDA problem.

We accordingly consider here the aggregation of outcomes A_0, A_1, and A_2 with multi-dimensional consequences, under a finite number of performance criteria $C_1,\ldots, C_m$. Our objective is to assess the chance fork with the outcomes A_1 and A_2 and their associated probabilities. To illustrate matters, we suppose that we have the following problem. A decision maker visits a number of dealers to select a car. Late in the afternoon he/she finds a car A_0 with an acceptable price and an acceptable reliability (for the time being, we ignore the other attributes of cars). The decision maker can now sign the contract, but he/she may also continue the search on the next day. On the basis of previous experience he/she estimates that there is a chance p_1 to find a cheaper, somewhat less reliable car A_1, and a chance $p_2 = 1 - p_1$ to find a somewhat more expensive and more reliable car A_2. If he/she continues the search, however, the deal with the car A_0 is over. How to decide now?

Under the assumption that the decision maker is only prepared to consider prices in the range between Dfl 20.000 (US$ 12.000) and Dfl 40.000 (US$ 24.000) and reliabilities between 95 % and 100 % (see Section 2.2) we obtain the assignment of grades which is shown in Table 2.2. Let us now suppose that the alternative cars have the prices and the reliabilities displayed in the performance tableau of Table 3.21, with the associated SMART impact grades between brackets.

Table 3.21 *Performance tableau of the alternative cars with SMART impact grades.*

Criteria	Alternative A_0	Alternative A_1	Alternative A_2
C_1: Price	22.500 (6)	21.200 (7)	25.000 (5)
C_2: Reliability	98.8 (6)	97.5 (5)	99.4 (7)

The pairwise-comparison matrices R_1 and R_2 under the price criterion and the reliability criterion are given by

$$R_1 = \begin{pmatrix} 1 & \frac{1}{2} & 2 \\ 2 & 1 & 4 \\ \frac{1}{2} & \frac{1}{4} & 1 \end{pmatrix} \text{ and } R_2 = \begin{pmatrix} 1 & 2 & \frac{1}{2} \\ \frac{1}{2} & 1 & \frac{1}{4} \\ 2 & 4 & 1 \end{pmatrix}$$

respectively. Thus, under the price criterion A_0 is twice as bad as A_1 and twice as good as A_2, see the first row in R_1, etc. Let us assume that the two criteria have equal weights. The normalized impact scores and the normalized terminal scores, calculated according to the procedure just described, may be found in Table 3.22, again with the associated SMART impact grades between brackets.

Table 3.22 *AHP scores and SMART grades of the alternative cars.*

Criteria		Alternative A_0	Alternative A_1	Alternative A_2
C_1: Price	0.5	0.29 (6)	0.57 (7)	0.14 (5)
C_2: Reliability	0.5	0.29 (6)	0.14 (5)	0.57 (7)
Terminal scores (final grades)		0.33 (6)	0.33 (6)	0.33 (6)

The above example is so symmetric that the decision maker must be indifferent between the sure outcome A_0 and the chance fork with the outcomes A_1 and A_2, whatever the method of analysis. When we use expected values to aggregate the outcomes, SMART does lead to indifference, both at the level of the impact grades and at the level of the final grades. With $p_1 = p_2 = 0.5$ we obtain

$$7p_1 + 5p_2 = 6,$$
$$5p_1 + 7p_2 = 6,$$
$$6p_1 + 6p_2 = 6.$$

Expected values of the AHP scores do not work so nicely. At the level of the impact scores of the alternatives we obtain

$$0.57\,p_1 + 0.14\,p_2 \neq 0.29,$$
$$0.14\,p_1 + 0.57\,p_2 \neq 0.29,$$

for $p_1 = p_2 = 0.5$. Since the Multiplicative AHP is an exponential version of SMART, however, we propose to use the probabilities p_1 and p_2 as exponents in a weighted geometric mean of scores in order to aggregate a chance fork. This yields indifference, because

$$0.57^{p_1} \times 0.14^{p_2} = 0.29,$$
$$0.14^{p_1} \times 0.57^{p_2} = 0.29,$$
$$0.33^{p_1} \times 0.33^{p_2} = 0.33,$$

when $p_1 = p_2 = 0.5$. Because the ability of human beings to estimate the probabilities of uncertain events is questionable we propose to consider p_1 and p_2 as coefficients expressing the relative importance of the alternatives in the chance fork.

We can now generalize the results. Suppose that we are concerned with a decision tree presenting the choice between the sure outcome A_0 on the one hand and the chance fork with the mutually exclusive outcomes A_j, $j = 1,\ldots,n$, and the associated probabilities p_j, $j = 1,\ldots,n$, on the other. Furthermore, the alternatives are to be judged under the criteria C_i, $i = 1,\ldots,m$, with corresponding criterion weights c_i. Working with SMART we compare the arithmetic means

$$\sum_{i=1}^{m} c_i g_{i0} \text{ and } \sum_{i=1}^{m}\sum_{j=1}^{n} c_i g_{ij} p_j, \tag{3.33}$$

where g_{ij} stands for the impact grade assigned to alternative A_j under criterion C_i, and we choose the arithmetic mean with the highest value. Obviously, the expressions in (3.33) have the form of an expected value (in fact expected values of logarithms). Working with the Multiplicative AHP we compare the geometric means

$$\prod_{i=1}^{m} s_{i0}^{c_i} \text{ and } \prod_{i=1}^{m}\prod_{j=1}^{n} s_{ij}^{c_i p_j}, \tag{3.34}$$

where s_{ij} designates the impact score of alternative A_j under criterion C_i, and we choose the geometric mean with the highest value. The expressions in (3.34) cannot be interpreted as expected values. That is the starting point of the considerations to follow.

The ability of human beings to estimate probabilities should not be over-estimated. Hence, a probabilistic model for the uncertainties in an actual decision problem may be a hazardous tool, regardless of whether the decision makers are requested to estimate the probabilities numerically or in verbal terms. We propose to follow a different route, however. The formulas in (3.33) and (3.34) show that, mathematically, the outcome probabilities and the criterion weights play the same role in the aggregation procedure. Chapter 6 presents a similar result in group aggregation procedures incorporating the relative power of the decision makers: the criterion weights and the power coefficients have identical positions in the aggregation formulas. Hence, we take the outcome probabilities to represent coefficients of importance: if A_1 is *somewhat more* likely to occur than A_2, we take it to be *somewhat more* important in the aggregate quantity representing the chance fork, etc. Thus, we associate the gradations of relative likelihood with the same scale values as the gradations of relative importance, so that

A_1 as likely as A_2,	odds 1:1,	$p_1 = 0.50$,	$p_2 = 0.50$,
A_1 somewhat more likely than A_2,	odds 2:1,	$p_1 = 0.66$,	$p_2 = 0.33$,
A_1 more likely than A_2,	odds 4:1,	$p_1 = 0.80$,	$p_2 = 0.20$,
A_1 much more likely than A_2,	odds 8:1,	$p_1 = 0.90$,	$p_2 = 0.10$,
A_1 vastly more likely than A_2,	odds 16:1,	$p_1 = 0.95$,	$p_2 = 0.05$.

The underlying idea is that comparative verbal quantifiers like *somewhat more, more, much more,* and *vastly more* have numerical values which do not depend on what we compare: the relative importance of the criteria, the relative power of the decision makers, or the relative likelihood of the outcomes. By this uniformity, comparative verbal quantifiers lubricate human communication. Imagine how irritating it would be if we always had to use numerical values instead of vague verbal terms with a quantitative connotation. The fact that comparative verbal quantifiers are usually well-understood is an argument in favour of the hypothesis that they have imprecise numerical values with a uniform validity.

With the above numerical scaling of relative likelihood we can ask the decision maker to estimate the outcome probabilities $p_1,\ldots, p_n$ via a method of pairwise comparisons. In the basic experiment we present a pair of possible outcomes to the decision maker, whereafter we can ask him/her to state whether they are *equally* likely or whether one of the two is *somewhat more, more, much more,* or *vastly more* likely to occur than the other. Next, we can convert the statements into elements of a pairwise-comparison matrix from which we eventually extract the outcome probabilities via the calculation of geometric row means.

The above method has not been considered in the literature, although the significance of verbal probability estimates has extensively been studied in the social sciences (see the author, 1997). In the reported experiments the respondents were usually requested to assign values to probability terms within the context of an uncertain situation or to select the most suitable probability term from a given list in order to characterize the situation. The authors discovered a high degree of context-dependent variability of the meaning of these terms, as well as a weak relationship between these terms and the actual probabilities in the given situations. Particularly Teigen (1988) reported an amazing degree of innumeracy and a serious lack of probabilistic thinking among the respondents when they were requested to estimate the probabilities of a number of mutually exclusive outcomes (this is precisely what a decision maker has to do in a chance fork). Brun and Teigen (1988), on the other hand, found that the interpretation variability does not pose a serious communication problem. Pairwise comparisons of outcome probabilities were not included in these studies, however. A method of pairwise comparisons can be used to estimate the ratios of the outcome probabilities. Thereafter, the results can be normalized to guarantee that the outcome probabilities sum to 1 or to 100%.

In summary, we aggregate the multi-dimensional consequences in a chance fork in such a manner that the results are compatible, whether we use the Multiplicative AHP with the geometric-mean aggregation rule, or its logarithmic counterpart SMART with the arithmetic-mean aggregation rule. The aggregation can be carried out in any order: first over the impact grades and scores and thereafter over the criteria, or vice versa (see the double summation in (3.33) and the double multiplication in (3.34)). We also made it plausible that verbal quantifiers like *somewhat more,..., vastly more* have imprecise numerical values which do not depend on the type of comparative judgement where they are used. Hence, they constitute a vehicle for communication, provided that we use the same quantifiers throughout the comparative assessment, regardless of whether we compare performance criteria or outcome probabilities.

REFERENCES TO CHAPTER 3

1. Bana e Costa, C.A., and Vansnick, J.C., "A Theoretical Framework for Measuring Attractiveness by a Category Based Evaluation Technique (MACBETH)". In J. Climaco (ed.), "*Multicriteria Analysis*", Springer, Berlin, 1997, pp. 15 – 24.

2. Barzilai, J., Cook, W.D., and Golany, B., "Consistent Weights for Judgement Matrices of the Relative Importance of Alternatives". *Operations Research Letters* 6, 131 – 134, 1987.

3. Barzilai, J., and Golany, B., "Deriving Weights from Pairwise Comparison Matrices: the Additive Case". *Operations Research Letters* 9, 407 – 410, 1990.

4. Barzilai, J., and Golany, B., "AHP Rank Reversal, Normalization, and Aggregation Rules". *INFOR* 32, 57 – 64, 1994.

5. Barzilai, J., and Lootsma, F.A., "Power Relations and Group Aggregation in the Multiplicative AHP and SMART". *Journal of Multi-Criteria Decision Analysis* 6, 155 – 165, 1997. In the same issue there are comments by O. Larichev (166), P. Korhonen (167 – 168), and L. Vargas (169 – 170), as well as a response by F.A. Lootsma and J. Barzilai (171 – 174).

6. Belton, V., and Gear, A.E., "On a Shortcoming of Saaty's Method of Analytical Hierarchies". *Omega* 11, 227 – 230, 1983.

7. Borcherding, K., and Winterfeldt, D. von, "The Effect of Varying Value Trees on Multi-attribute Evaluations". *Acta Psychologica* 68, 153 – 170, 1988.

8. Bradley, R.A., and Terry, M.E., "The Rank Analysis of Incomplete Block Designs. I. The Method of Paired Comparisons". *Biometrika* 39, 324 – 345, 1952.

9. Brun, W., and Teigen, K.H., "Verbal Probabilities: Ambiguous, Context-Dependent, or Both?". *Organizational Behavior and Human Decision Processes* 41, 390 – 404, 1988.

10. Budescu, D.V., Crouch, B.D., and Morera, O.F., "A Multi-Criteria Comparison of Response Scales and Scaling Methods in the AHP". In W.C. Wedley (ed.), *Proceedings of the Fourth International Symposium on the AHP.* Simon Fraser University, Burnaby, B.C., Canada, 1996, pp. 280 – 291.

11. Cogger, K.O., and Yu, P.L., "Eigenweight Vectors and Least-Distance Approximations for Revealed Preferences in Pairwise Weight Ratios". *Journal of Optimization Theory and Applications* 46, 483 – 491, 1985.

12. Crawford, G., and Williams, C., "A Note on the Analysis of Subjective Judgement Matrices". *Journal of Mathematical Psychology* 29, 387 – 405, 1985.

13. David, H.A., *"The Method of Paired Comparisons".* Griffin, London, 1963.

14. Dyer, J.S., "Remarks on the Analytic Hierachy Process". *Management Science* 36, 249 – 258, 1990. In the same issue there are apologies by T.L. Saaty (259 – 268), P.T. Harker and L. Vargas (269 – 273), and a further clarification by J.S. Dyer (274 – 275).

15. French, S., *"Decision Theory, an Introduction to the Mathematics of Rationality"*. Ellis Horwood, Chichester, 1988.

16. Johnson, C.R., Beine, W.B., and Wang, T.J., "Right-Left Asymmetry in an Eigenvector Ranking Procedure". *Journal of Mathematical Psychology* 19, 61 – 64, 1979.

17. Keeney, R., and Raiffa, H., *"Decisions with Multiple Objectives: Preferences and Value Trade-offs"*. Wiley, New York, 1976.

18. Lootsma, F.A., "Saaty's Priority Theory and the Nomination of a Senior Professor in Operations Research". *European Journal of Operational Research* 4, 380 – 388, 1980.

19. Lootsma, F.A., "Modélisation du Jugement Humain dans l'Analyse Multicritère au Moyen de Comparaisons par Paires". *RAIRO/Recherche Opérationnelle* 21, 241 – 257, 1987.

20. Lootsma, F.A., "Scale Sensitivity in the Multiplicative AHP and SMART". *Journal of Multi-Criteria Decision Analysis* 2, 87 – 110, 1993.

21. Lootsma, F.A., "Multi-Criteria Decision Analysis in a Decision Tree". *European Journal of Operational Research* 101, 442 – 451, 1997.

22. Lootsma, F.A., Boonekamp, P.G.M., Cooke, R.M., and Oostvoorn, F. van, "Choice of a Long-Term Strategy for the National Electricity Supply via Scenario Analysis and Multi-Criteria Analysis". *European Journal of Operational Research* 48, 189 – 203, 1990.

23. Mintzberg, H., *"Power in and around Organizations"*. Prentice-Hall, Englewood Cliffs, N.J., 1983.

24. Pöyhönen, M., and Hämäläinen, R.P., "Notes on the Weighting Biases in Value Trees". To appear in the *Journal of Behavioural Decision Making*, 1998.

25. Ramanathan, R., "A Note on the Use of Goal Programming for the Multiplicative AHP". *Journal of Multi-Criteria Decision Analysis* 6, 296 – 307, 1997.

26. Saaty, T.L., "A Scaling Method for Priorities in Hierarchical Structures". *Journal of Mathematical Psychology* 15, 234 – 281, 1977.

27. Saaty, T.L., *"The Analytic Hierarchy Process, Planning, Priority Setting, and Resource Allocation"*. McGraw-Hill, New York, 1980.

28. Saaty, T.L., and Vargas, L.G., "Inconsistency and Rank Preservation". *Journal of Mathematical Psychology* 28, 205 – 214, 1984.

29. Stewart, T.J., "A Critical Survey on the Status of Multi-Criteria Decision Making Theory and Practice". *Omega* 20, 569 – 586, 1992.

30. Takeda, E., Cogger, K.O., and Yu, P.L., "Estimating Criterion Weights using Eigenvectors: a Comparative Study". *Omega* 20, 569 – 586, 1987.

31. Teigen, K.H., "When are Low-Probability Events judged to be Probable?. Effects of Outcome-Set Characteristics on Verbal Probability Estimates". *Acta Psychologica* 67, 157 – 174, 1988.

32. Thurstone, L.L., "The Method of Paired Comparisons for Social Values". *Journal of Abnormal Psychology* 21, 384 – 400, 1927.

33. Torgerson, W.S., "Distances and Ratios in Psycho-Physical Scaling". *Acta Psychologica* XIX, 201 – 205, 1961.

34. Triantaphyllou, E., Lootsma, F.A., Pardalos, P.M., and Mann, S.H., "On the Evaluation and Application of Different Scales for Quantifying Pairwise Comparisons in Fuzzy Sets". *Journal of Multi-Criteria Decision Analysis* 3, 133 – 155, 1994.

35. Winterfeldt, D. von, and Edwards, W., "*Decision Analysis and Behavioral Research*". Cambridge University Press, Cambridge, UK, 1986.

CHAPTER 4

SCALE SENSITIVITY AND RANK PRESERVATION

In the previous chapters we have extensively used a geometric scale in order to model the gradations of comparative human judgement. Geometric progression seems to be reasonable but the progression factor 2 established on the basis of the categorization of time, space, light, and sound, remains questionable. The present chapter is therefore concerned with a one-parametric class of geometric scales. We analyze the behaviour of the AHP and SMART when the so-called scale parameter varies over a positive range of values. Theoretical arguments and numerical experiments with small and large problems show that the scale sensitivity of the terminal scores is low whereas the rank order remains unchanged.

4.1 A ONE-PARAMETRIC CLASS OF GEOMETRIC SCALES

Let us return to the evaluation of cars under the consumer price criterion (Section 2.1) and let us reconsider the geometric progression of the echelons on the range (P_{min}, P_{max}) of acceptable prices. The behaviour of the echelons has been expressed by the formula

$$e_{\nu} = (1+\varepsilon)^{\nu} e_0,$$

where the parameter ν designates the order of magnitude. Instead of setting the progression factor $(1 + \varepsilon)$ to the value 2, we generalize the geometric behaviour of the echelons by setting

$$e_\nu = 2^{\gamma\nu} e_0.$$

The symbol γ stands for the scale parameter. Obviously, $\gamma = {}^2\log(1 + \varepsilon)$. In the previous chapters we have been working with $\gamma = 1$. We shall now consider variations of γ in the neighbourhood, under the assumption that there are again seven echelons corresponding to four major and three threshold gradations of comparative judgement (cheap, somewhat more, more, and much more expensive, and the gradations between them). The associated price levels can accordingly be written as

$$P_\nu = P_{\min} + 2^{\gamma\nu} e_0, \quad \nu = 0, 1, \ldots, 5,$$

$$P_{\max} = P_{\min} + 2^{6\gamma} e_0,$$

which implies

$$e_0 = \frac{P_{\max} - P_{\min}}{2^{6\gamma}}.$$

We now "cover" the range of acceptable prices by the geometric sequence of points

$$P_\nu = P_{\min} + (P_{\max} - P_{\min}) \times \frac{2^{\gamma\nu}}{2^{6\gamma}}, \quad \nu = 0, 1, \ldots, 6. \tag{4.1}$$

The order of magnitude ν and the corresponding price level P_ν are connected by the relationship

$$\nu = \frac{1}{\gamma} \times {}^2\log\left(\frac{P_\nu - P_{\min}}{P_{\max} - P_{\min}} \times 2^{6\gamma}\right). \tag{4.2}$$

The formulas (4.1) and (4.2) are generalizations of the formulas (2.1) and (2.2). Let us now consider two alternative cars A_j and A_k which belong to the price categories represented by the price levels

$$P_{\nu_j} \text{ and } P_{\nu_k}$$

respectively. The relative preference for A_j with respect to A_k is again expressed by the inverse ratio of the price increments above $P_{\min}$ so that it is given by

$$\frac{P_{\nu_k} - P_{\min}}{P_{\nu_j} - P_{\min}} = 2^{\gamma(\nu_k - \nu_j)}. \tag{4.3}$$

This is a straightforward generalization of formula (2.3). Comparing the cars in the price categories P_0 and P_2 we identify the ratio $2^{2\gamma}:1$ with weak preference. Similarly, the ratio $2^{4\gamma}:1$ can be identified with strict preference, and the ratio $2^{6\gamma}:1$ with strong preference. Using formula (2.6), the relationship between the grades and the orders of magnitude of the judgemental categories, we can rewrite (4.3) in the form

$$\frac{P_{v_k} - P_{\min}}{P_{v_j} - P_{\min}} = 2^{\gamma(g_j - g_k)}.$$

The gradations of relative preference under the criteria of reliability and maximum speed can be quantified in a similar manner. In order to illustrate the results so far we reconsider the assignment of grades in Table 2.2 and we recalculate the corresponding levels for two different values of the scale parameter. Table 4.1 exhibits the assignment when γ is set to 0.72 so that the progression factor is given by exp(0.5) = 1.65. In Table 4.2 the parameter γ has been set to 1.44, which implies that the progression factor is exp(1.0) = 2.72.

Table 4.1 *Assignment of grades within predetermined ranges. The geometric scale has the scale-parameter value* $\gamma = 0.72$ *and henceforth the progression factor* 1.65.

Price	Reliability	Max. speed	Performance	Grade	Symbolic scale
21,000	99.75	220	Excellent	10	+++
21,500	99.60	190	Good/excellent	9	++
22,500	99.30	170	Good	8	+
24,500	98.90	160	Fair/good	7	o
27,500	98.20	150	Fair	6	-
32,000	97.00	147	Unsatisfactory/fair	5	- -
40,000	95.00	144	Unsatisfactory	4	- - -

Table 4.2 *Assignment of grades within predetermined ranges. The geometric scale has the scale-parameter value* $\gamma = 1.44$ *and henceforth the progression factor* 2.72.

Price	Reliability	Max. speed	Performance	Grade	Symbolic scale
20,000	99.99	220	Excellent	10	+++
20,100	99.97	170	Good/excellent	9	++
20,400	99.91	150	Good	8	+
21,000	99.75	144	Fair/good	7	o
22,700	99.30	141.5	Fair	6	-
27,400	98.20	140.5	Unsatisfactory/fair	5	- -
40,000	95.00	140.2	Unsatisfactory	4	- - -

In the Multiplicative AHP we can now readily generalize the assignment of numerical values to the gradations of relative preference or comparative judgement (Section 3.1). Thus,

$$r_{jk} = 2^{\gamma(g_j - g_k)} = 2^{\gamma\delta_{jk}}, \tag{4.4}$$

$$^2\log r_{jk} = \gamma\delta_{jk}, \tag{4.5}$$

where δ_{jk} represents the difference of grades g_j - g_k (see the formulas (3.2) - (3.6)). In what follows we shall equivalently take δ_{jk} to designate the gradation which has been selected by the decision maker in order to express his/her relative preference for A_j with respect to A_k, and we shall accordingly refer to δ_{jk} as the gradation index.

Table 4.3 shows the numerical values assigned to the gradations of comparative judgement when we use a variety of geometric scales. It is easy to see that Table 4.3 (with gradations of desirability) generalizes Table 3.1 (with gradations of expensiveness) and Table 3.2 (with gradations of reliability).

Table 4.3 *Scale values assigned to gradations of comparative judgement on a variety of geometric scales (Multiplicative AHP).*

Comparative judgement of A_j with respect to A_k	Gradation index δ_{jk}	Scale values r_{jk} $\gamma = 0.72$	$\gamma = 1.00$	$\gamma = 1.44$
A_j much more desirable than A_k	6	20.1	64	403
A_j more desirable than A_k	4	7.4	16	55
A_j somewhat more desirable than A_k	2	2.7	4	7
A_j as desirable as A_k	0	1	1	1
A_j somewhat less desirable than A_k	-2	0.37	0.25	0.14
A_j less desirable than A_k	-4	0.135	0.063	0.018
A_j much less desirable than A_k	-6	0.050	0.016	0.002

So, basically, we vary the price levels, the reliability levels, etc., corresponding to the grades. This can also be seen in the Figures 2.2 - 2.5. The curve representing the relationship between the echelons and the orders of magnitude becomes more concave when γ increases. The grades on the vertical axis remain unchanged, the corresponding echelons move along the horizontal axis with variations of γ.

In the remainder of this chapter we shall therefore concentrate on the Multiplicative AHP when we analyze the scale sensitivity of the AHP via theoretical considerations and numerical experiments (see also the author, 1993). We shall use the gradation indexes in order to determine the scale values, whereafter we calculate the impact scores and the terminal scores of the alternatives with a variety of geometric scales.

4.2 RANK PRESERVATION OF THE SCORES

The theoretical sensitivity analysis proceeds in three steps. We first demonstrate that the rank order of the impact scores is preserved under scale variations within the one-parametric class of geometric scales. Next, we show that we can use a uniform scale in order to model the relative importance of the criteria. Thereafter it will be easy to see that the rank order of the terminal scores is preserved as well.

Rank preservation of the impact scores. We consider the alternatives under an unspecified criterion. Let us again take the symbol r_{jkd} to stand for the numerical value assigned to the judgemental statement expressed by decision maker d in the experiment where he/she compares alternative A_j with respect to A_k. Thus, r_{jkd} is an estimate of the ratio V_j/V_k of the subjective values of the respective alternatives under consideration, whereas

$$^2\log r_{jkd} = \gamma\delta_{jkd},$$

with the gradation index δ_{jkd}. We approximate the subjective values of the alternatives via the unconstrained minimization of the sum of squares

$$\sum_{j>k}\sum_{d\in D_{jk}}\left(^2\log r_{jkd} - {}^2\log \upsilon_j + {}^2\log \upsilon_k\right)^2. \tag{4.6}$$

Introducing the variables $w_j = {}^2\log \upsilon_j$ we rewrite the function (4.6) in the form

$$\sum_{j>k}\sum_{d\in D_{jk}}\left(\gamma\delta_{jkd} - w_j + w_k\right)^2. \tag{4.7}$$

A minimum solution of (4.7) is given by a solution to the normal equations obtained by setting the first-order derivatives to zero. Hence,

$$\sum_{\substack{k=1\\k\neq j}}^{n}\sum_{d\in D_{jk}}\left(\gamma\delta_{jkd} - w_j + w_k\right) = 0,$$

$$w_j\sum_{\substack{k=1\\k\neq j}}^{n} N_{jk} - \sum_{\substack{k=1\\k\neq j}}^{n} N_{jk}w_k = \gamma\sum_{\substack{k=1\\k\neq j}}^{n}\sum_{d\in D_{jk}}\delta_{jkd},\; j = 1,\ldots,n, \tag{4.8}$$

where N_{jk} denotes again, like in Section 3.2, the cardinality of the set D_{jk} of decision makers who judged A_j with respect to A_k. Because the scale parameter γ appears linearly in the right-hand side only, we can write the components of a general solution $w(\gamma)$ to the normal equations (4.8) in the form

$$w_j(\gamma) = \gamma w_j(1) + \theta,\ j = 1, \ldots, n.$$

The vector $w(1)$ represents an arbitrary solution to the normal equations for $\gamma = 1$, and θ stands for the additive degree of freedom in the system. Throughout the analysis the vector $w(1)$ will remain unchanged. The components of the normalized solution $s(\gamma)$ are accordingly given by the expression

$$s_j(\gamma) = \frac{2^{\gamma w_j(1)}}{\sum_{j=1}^{n} 2^{\gamma w_j(1)}},\ j = 1, \ldots, n.$$

Normalization eliminates the additive degree of freedom θ. The components of $s(\gamma)$ are the impact scores of the alternatives under the unspecified criterion. For any pair of alternatives A_j and A_k the ratio of the impact scores is

$$\frac{s_j(\gamma)}{s_k(\gamma)} = 2^{\gamma(w_j(1) - w_k(1))}. \tag{4.9}$$

The positive scale parameter γ does not affect the rank order of the alternatives because the ratio (4.9) is greater (smaller) than unity when the fixed quantity $w_j(1)$ - $w_k(1)$ is positive (negative). In the sections to follow we will show that reasonable variations of γ hardly affect the numerical values of the impact scores.

A uniform scale for the relative importance of the criteria. Let us again carry out the hypothetical experiment of Section 2.7 in order to establish a scale of values for the relative importance of two criteria C_f and C_s. Thus, we ask the decision maker to consider two real or imaginary alternatives A_j and A_k such that his/her preference for A_j versus A_k under the first criterion C_f equals his/her preference for A_k versus A_j under the second criterion C_s. The relative preference for A_j with respect to A_k under C_f and C_s is represented by the ratios

$$2^{\gamma\delta_{jk}} \text{ and } 2^{-\gamma\delta_{jk}}$$

respectively, where δ_{jk} stands for the gradation index corresponding to the decision maker's judgement. We also ask him/her to judge the two alternatives under the two criteria simultaneously. Let us suppose that his/her relative preference may be encoded as

$$2^{\gamma\vartheta_{jk}}$$

where ϑ_{jk} designates the gradation that has been chosen by the decision maker to express his/her judgement. If we take the symbol ω to denote the relative importance of the two criteria (the ratio of the corresponding criterion weights) we have, by the geometric-mean aggregation rule in the Multiplicative AHP so that we consider an arithmetic mean of the exponents, the following relationship between the relative preferences under C_f and C_s individually and the relative preference under the two criteria simultaneously:

$$\left(\gamma \frac{\omega}{\omega+1}\delta_{jk}\right)+\left(-\gamma \frac{1}{\omega+1}\delta_{jk}\right)=\gamma\theta_{jk}, \tag{4.10}$$

which eventually leads to

$$\omega=\frac{\delta_{jk}+\vartheta_{jk}}{\delta_{jk}-\vartheta_{jk}}.$$

Hence, the effect of the ratio ω in the aggregation procedure does not depend on the scale parameter γ. This is the crucial observation. In the sensitivity analysis of the sections to follow we therefore use the geometric scale with the progression factor $\sqrt{2}$ (and henceforth the scale parameter $\gamma = 0.5$) to quantify the gradations of relative importance, the same scale that we used throughout the Chapters 2 and 3. The criterion weights do not therefore depend on the scale parameter γ employed in the theoretical analysis of the present section and in the numerical experiments of the sections to follow.

Note that substitution of $\delta_{jk} = 8$ and $\vartheta_{jk} = 7$ into (4.10) yields $\omega = 15$, a value that we also found by formula (2.12). The two cases modelled by (2.12) and (4.10) are identical: the relative importance of the first criterion with respect to the second one is *vastly higher*, because the decision maker's *very strong* preference for A_j with respect to A_k under the first criterion almost completely overrules the equally strong but inverse preference under the second criterion. On the basis of this result we introduced the uniform scale for the gradations of relative importance, a geometric scale between $^1/16$ and 16, with echelons corresponding to *somewhat more, more, much more,* and *vastly more* importance, and the threshold gradations between them.

Rank preservation of the terminal scores. The relative preference for alternative A_j with respect to A_k under all criteria simultaneously, in other words the ratio of the terminal scores, is given by the weighted geometric mean of the relative preferences under the individual criteria. It can accordingly be written in the form

$$\frac{t_j(\gamma)}{t_k(\gamma)}=\prod_{i=1}^{m}\left(\frac{s_{ij}(\gamma)}{s_{ik}(\gamma)}\right)^{c_i}=2^{\gamma z_{jk}},\ \text{with}\ z_{jk}=\sum_{i=1}^{m}c_i\left[w_{ij}(1)-w_{ik}(1)\right]. \tag{4.11}$$

We have obviously used the symbols in formula (4.9) with an extra index i to designate the criterion under consideration. Under variations of the scale parameter γ over a positive range of values, the ratio (4.11) remains greater (smaller) than unity whenever the fixed quantity z_{jk} is positive (negative). The highest terminal score remains the highest one, etc. Thus, the rank order of the terminal scores is preserved under scale variations within the one-parametric class of geometric scales. Furthermore, the analysis of the scale sensitivity can be simplified considerably. When the pairwise comparisons are complete, there is an explicit solution to the normal equations. Taking

$$r_{ijk} = 2^{\gamma\delta_{ijk}},$$

with the gradation index δ_{ijk} to encode the decision maker's judgement, and ignoring the normalization, we can write the terminal scores in the form

$$t_j(\gamma) = \prod_{i=1}^{m} s_{ij}(\gamma)^{c_i} = \sqrt[n]{\prod_{i=1}^{m}\prod_{k=1}^{n} r_{ijk}^{\ c_i}} = 2^{\gamma b_j},$$

$$\text{with } b_j = \frac{1}{n}\sum_{k=1}^{n}\sum_{i=1}^{m} c_i\delta_{ijk} = \frac{1}{n}\sum_{k=1}^{n} a_{jk}, \text{ and } a_{jk} = \sum_{i=1}^{m} c_i\delta_{ijk}.$$

The term a_{jk} is the (j, k)-th component of the aggregate pairwise-comparison matrix (in logarithmic form) obtained by the aggregation of the pairwise-comparison matrices under the individual criteria. Hence, the terminal scores can easily be obtained via the calculation of the arithmetic row means of the aggregate matrix. We illustrate this via two examples. First we consider the nomination procedure discussed in Section 3.4. Table 3.7 shows the differences of grades or, equivalently, the gradation indexes, as well as the aggregate matrix and its arithmetic row means. In Table 4.4 we exhibit the AHP terminal scores for various values of the scale parameter γ. It will be clear that the scale sensitivity is low. There are considerable variations in the lengths of the scales (see also Table 4.3) but the AHP terminal scores are hardly affected. Next, we consider the example of Belton and Gear (1983) with the data displayed in Table 3.17. The evaluation of the alternatives under the respective criteria stretches over the full length of the scale, from -8 to 8. Table 4.5 shows the aggregate matrix and the AHP terminal scores for different values of the scale parameter. Obviously, the scale sensitivity is low again.

Table 4.4 *Evaluation of three candidates under seven performance criteria (Table 3.7) for various values of the scale parameter γ. The arithmetic row means of the aggregate pairwise-comparison matrix can easily be used to calculate the AHP terminal scores.*

	A OR pract.	*B* teaching exp.	*C* active publicist	arithm. row means	AHP term. sc. $\gamma = 0.72$	AHP term. sc. $\gamma = 1.00$	AHP term. sc. $\gamma = 1.44$
A	0	0.309	0.382	0.230	0.373	0.388	0.414
B	-0.309	0	0.073	-0.079	0.319	0.314	0.304
C	-0.382	-0.073	0	-0.151	0.308	0.298	0.282

Table 4.5 *Example of Belton and Gear* (1983) *with rescaled data* (*Table* 3.17, *which shows differences of grades or, equivalently, gradation indexes*). *The arithmetic row means of the aggregate pairwise-comparison matrix can easily be used to calculate the AHP terminal scores for various values of the scale parameter* γ.

	A_1	A_2	A_3	A_4	arithm. row means	AHP term. sc. $\gamma = 0.72$	AHP term. sc. $\gamma = 1.00$	AHP term. sc. $\gamma = 1.44$
A_1	0	-0.333	5	-0.333	1.0835	0.290	0.281	0.263
A_2	0.333	0	5.333	0	1.4165	0.343	0.355	0.367
A_3	-5	-5.333	0	-5.333	-3.9165	0.024	0.009	0.002
A_4	0.333	0	5.333	0	1.4165	0.343	0.355	0.367

4.3 EVALUATION OF ENERGY R & D AREAS

The energy crises of 1973 and 1979 prompted many governments to provide substantial budgets for Energy Research and Development. In 1980 the Government of the Netherlands set up the Energy Research Council (*Raad voor het Energie Onderzoek*) with the task (a) to give advice about all matters concerning the national Energy R & D, and (b) to present every year a medium-term plan for a national Energy R & D policy. A major item in this plan, which is also the subject of the present case study, would be a proposed allocation of the available budget to various areas for Energy R & D. The proposed budget allocation would eventually be submitted to the Ministry of Economic Affairs which could either accept or reject the recommendations, all or in part, since the Minister would actually be the accountable decision maker.

The Council would obviously be concerned with Energy R & D in domains of interest where the socio-economic benefits emerge in later years. Although it might be possible to assess the costs of relevant R & D in the respective technological areas, it would be impossible to put a monetary value on the potential future benefits. The Council members foresaw that they could only assess the relative merits of Energy R & D in these areas on the basis of comparative human judgement.

The Energy Research Council had 18 members representing major research institutes, universities, industries, trade unions, consumer organisations, environmental agencies, and lower public authorities. Because of this heterogeneity the Council members would view the budget-allocation problem from rather different angles. In the end, however, some kind of consensus should emerge. The Council therefore set up a working group with the task to review the objectives and the criteria for the evaluation of the Energy R & D areas. Moreover, the working group was expected to present an appropriate methodology for the budget-allocation problem. When the Council was installed in 1980, the Minister of

Economic Affairs mentioned as the leading objectives for a national Energy R & D policy: (a) reduction of the energy demand, and (b) diversification of the energy resources. Bearing this in mind the working group analyzed the obstacles for such a policy:

- The Netherlands are heavily dependent on oil imports and domestic natural gas so that energy production and energy consumption have a low flexibility. The national economy is rather sensitive to supply and price fluctuations. The large oil imports have a substantial and uncertain impact on the balance of payments. High energy prices may affect the competitive position of the national energy-intensive industry. This can have repercussions on the economic activity and henceforth on employment.

- The long-term development of the energy prices and the development of the price ratios amongst various energy carriers are very uncertain. This hampers the adjustment of Energy R & D (which should be ahead of price developments) to new opportunities and threats.

- The introduction of new energy-producing systems is rather slow. This is due to various factors: the high cost of capital investment, a growing concern about environmental damage, strong public opposition against nuclear energy, an abundant supply of natural gas, and obstruction by lower public authorities. The introduction of energy-saving equipment like the heatpump is sometimes delayed because the existing equipment has not yet been fully depreciated.

- Small and medium-size industrial companies accept new developments too slowly, because of insufficient mutual cooperation, limited international orientation, lack of financial resources, and limited knowledge of the energy market. They also seem to be unfamiliar with the knowledge and expertise offered by government agencies and research institutes.

Considering these obstacles the working group formulated five criteria to judge various Energy R & D areas on the basis of their possible contributions to the establishment of a balanced national energy economy and to the consolidation of a competitive national industry, both at home and in foreign markets. In a pilot session the Council assessed the proposed criteria via a method of pairwise comparisons. On the basis of the results the Council improved the formulation and introduced a new criterion. Hence, the assessment of the Energy R & D efforts in the respective areas has eventually been carried out on the basis of their possible contribution to:

1. The diversification and the security of the national energy system. This will minimize the strategic risk which is due to the dependence on a small number of energy carriers.

2. The energy efficiency in the sectors of winning, conversion, distribution, and/or consumption. The increased efficiency may be achieved via the development of new technologies, a reduction of the energy demand, and/or the utilization of residues.

3. The social acceptability of new energy systems, which can be redesigned in order to satisfy the increasingly higher demands of public health, safety, and the environment.

4. The creation of innovative industrial activities, given the national energy position and the characteristics of the national industry.

5. The enhancement of long-term energy management via the delayed depletion of finite, high-quality energy sources, the utilization of sustainable technologies, and the management of uncertainties in the development of energy prices.

6. The advancement of high-level scientific activities via the exploitation of the specific features of the national R & D and/or the creation of spearheads in it.

Initially, the Council classified the Energy R & D activities according to the proposals of the International Energy Agency. The resulting domains were too large and too vaguely defined for an assessment via MCDA, however. This prompted the Council to refine the classification and to specify technological areas for Energy R & D which were relevant in the Netherlands and which could also be analysed via MCDA, at least at the level of support which is required for a relevant effort. Table 4.6 shows the 9 domains of interest and the 23 areas which were eventually assessed by 9 Council members in the working group. A complete description of the assessment has been reported by Légrády et al. (1984) and the author et al. (1986). Here we only consider the experiment in order to illustrate (a) the effectiveness of a pairwise-comparison method in MCDA with a large number of alternatives, and (b) the effects of scale variations within the one-parametric class of geometric scales.

Of course, it is practically impossible to carry out the assessment of so many alternatives via pairwise comparisons. With 6 criteria and 23 alternatives, the number of pairs to be considered would be $6 \times (0.5 \times 23 \times 22) = 1518$, and with 9 members in the working group, the analysts would have to input $9 \times 1518 = 13662$ entries. The tasks would be boring, and input errors could easily be made. Moreover, not every member was sufficiently familiar with the alternatives to judge all of them.

The members of the working group therefore selected the alternatives which they knew sufficiently well. The group, however, "covered" the whole field. The verbal statements of the group filled the cells of the pairwise-comparison matrices reasonably well, although each member assessed not more than 12 alternatives.

Figure 4.1 shows the typical form of the questionnaire submitted to the group members for the assessment of the criteria. In addition, each member received a questionnaire for the pairwise comparison of the alternatives which he/she had previously selected, so that it was not necessary to sort out the pairs to be judged. Eventually, however, the analysis would only produce a group opinion about the alternatives. The opinions of the individual members could not be retrieved.

Table 4.6 *Domains of interest and technological areas for Energy R & D, with the level of support which is required for a relevant effort.*

Domain of interest, IEA classification	Technological area for Energy R & D	Required support in millions of Dfl
Energy Saving	1. in Industry	40
	2. in Houses and Offices	30
	3. in Traffic and Transport	10
	4. Other (incl. heatpumps)	10
Oil and Gas	5. Oil and Natural Gas	3
Coal	6. Winning	10
	7. Combustion	35
	8. Gassification	25
	9. Residues	25
Nuclear Energy	10. Thermal Reactors	50
	11. Fission Cycle	25
	12. Breeders	50
	13. Fusion	25
Solar Energy	14. Thermal	10
	15. Electric	10
Wind Energy	16. Decentralized	10
	17. Centralized	20
Biomass Energy	18. Biomass	10
Geothermal Energy	19. Geothermal	10
Supporting Technology	20. Power Production	5
	21. Thermal Energy Storage	10
	22. Systems Analysis	10
	23. Relay Systems	5
Budget required for government-supported Energy R & D		438

We now first present the criterion weights and the terminal scores obtained on the basis of the analysis in Section 4.2 and with the original data. Table 4.7 exhibits the results. There is a uniform scale for the criteria but a variety of geometric scales for the alternatives, whereas the aggregation is carried out via geometric-mean calculations. Obviously, the terminal scores are hardly affected by the scale parameter. They have the same rank order.

The method employed in the experiment of 1983, however, was a variant of the AHP, with several geometric scales to be used simultaneously for the evaluation of the alternatives and the criteria, and with the arithmetic-mean aggregation rule to process ratio information. The resulting terminal scores (Légrády et al., 1984; Lootsma, 1987) are exhibited in Table 4.8. Their numerical values did not strongly vary with the length of the scale, but their rank order was scale-dependent, a phenomenon which worried the Council. In 1990 we implemented a modified aggregation procedure for ratio information: we replaced the arithmetic-mean calculations by geometric-mean calculations. This hardly affected the terminal scores, as Table 4.9 shows, and in some projects it reduced the number of rank reversals which were due to scale variations.

```
==========+
  FOR     |   Compare the impact of the following pairs of CRITERIA and
 OFFICE   |   mark the appropriate box.
USE ONLY  |
          |                   |    has     | has ABOUT |    has     |
          |                   |  SOMEWHAT  |   EQUAL   |  SOMEWHAT  |
   +----+ |        |   has    |   HIGHER   | impact as |   LOWER    |   has    |
1-4|   4| |        |   MUCH   |   impact   +---+   +---+   impact   |   MUCH   |
   +----+ |        |  HIGHER  |    than        |   |        than    |  LOWER   |
          |        |  impact  +------------+   |   |   +-----------+  impact  |
          |        |                       |   |   |   |                      |
56 78     |        +-----------------+     |   |   |   |     +----------------+
|| ||     |                          |     |   |   |   |     |
YY YY     |                          Y     Y   Y   Y   Y     Y
     +===+                           +----+---+---+----+----+
3- 2|    |   Social Acceptabili      |    |   |   |    |    | Energy Efficiency
     +===+                           +----+---+---+----+----+
2- 5|    |   Energy Efficiency       |    |   |   |    |    | Long-term Contribu
     +===+                           +----+---+---+----+----+
4- 2|    |   Innovation-based I      |    |   |   |    |    | Energy Efficiency
     +===+                           +----+---+---+----+----+
4- 3|    |   Innovation-based I      |    |   |   |    |    | Social Acceptabili
     +===+                           +----+---+---+----+----+
2- 6|    |   Energy Efficiency       |    |   |   |    |    | High-level Scienti
     +===+                           +----+---+---+----+----+
1- 2|    |   Security of Energy      |    |   |   |    |    | Energy Efficiency
     +===+                           +----+---+---+----+----+
5- 6|    |   Long-term Contribu      |    |   |   |    |    | High-level Scienti
     +===+                           +----+---+---+----+----+
5- 3|    |   Long-term Contribu      |    |   |   |    |    | Social Acceptabili
     +===+                           +----+---+---+----+----+
1- 6|    |   Security of Energy      |    |   |   |    |    | High-level Scienti
     +===+                           +----+---+---+----+----+
4- 6|    |   Innovation-based I      |    |   |   |    |    | High-level Scienti
     +===+                           +----+---+---+----+----+
1- 5|    |   Security of Energy      |    |   |   |    |    | Long-term Contribu
     +===+                           +----+---+---+----+----+
3- 1|    |   Social Acceptabili      |    |   |   |    |    | Security of Energy
     +===+                           +----+---+---+----+----+
1- 4|    |   Security of Energy      |    |   |   |    |    | Innovation-based I
     +===+                           +----+---+---+----+----+
5- 4|    |   Long-term Contribu      |    |   |   |    |    | Innovation-based I
     +===+                           +----+---+---+----+----+
3- 6|    |   Social Acceptabili      |    |   |   |    |    | High-level Scienti
     +===+                           +----+---+---+----+----+
```

Figure 4.1 *Pairs of criteria to be judged by the Energy Research Council. The member in question was invited to express his/her comparative judgement by a cross or tick in the appropriate box or on the vertical line separating two adjacent boxes. The positions of the crosses or ticks, designating the selected gradations of comparative judgement, were converted into numerical values on a variety of geometric scales. Today, the positions would only be converted into numerical values on the geometric scale with* $\gamma = 0.50$.

Table 4.7 *Weights of the criteria and AHP terminal scores of the alternative Energy R & D technologies assessed on the basis of their possible contributions to the predominant goals of a national Energy R & D policy. The procedure since* 1992: *there is a uniform scale for the gradations of relative importance but the gradations of relative performance are converted into numerical values on three different geometric scales. Aggregation of ratio information via the calculation of geometric means.*

Criteria Potential to Increase	Uniform scale for relative importance of criteria $\gamma = 0.50$			Rank order
Security of Energy Supply		0.224		2
Energy Efficiency		0.142		4
Social Acceptability		0.109		6
Innovative Industry		0.184		3
Long-Term Energy Management		0.226		1
High-Level Scientific Activity		0.114		5
Alternatives Energy R & D technologies	**Varying scales for relative performance of alternatives**			**Rank order**
	$\gamma = 0.72$	$\gamma = 1.00$	$\gamma = 1.44$	
Energy Saving in Industry	0.054	0.058	0.065	4
Energy Saving in Houses, Offices	0.057	0.062	0.071	2
Energy Saving in Transport	0.037	0.034	0.031	18
Energy Saving in Other Areas	0.052	0.055	0.060	5
Oil and Natural Gas	0.042	0.041	0.039	11
Coal Winning	0.044	0.043	0.043	9
Coal Combustion	0.043	0.042	0.041	10
Coal Gassification	0.055	0.060	0.067	3
Coal Residues	0.040	0.038	0.035	13
Uranium Thermal Reactor	0.041	0.040	0.038	12
Uranium Fission Cycle	0.047	0.048	0.049	7
Uranium Breeder	0.066	0.077	0.096	1
Uranium Nuclear Fusion	0.036	0.033	0.029	21
Solar Energy, Thermal	0.038	0.035	0.031	17
Solar Energy, Electrical	0.046	0.047	0.048	8
Wind Energy, Decentralized	0.037	0.034	0.030	19
Wind Energy, Centralized	0.040	0.038	0.035	14
Biomass Energy	0.039	0.037	0.034	16
Geothermal Energy	0.030	0.026	0.021	23
Electricity Production	0.036	0.034	0.030	20
Thermal Energy Storage	0.040	0.038	0.035	15
Systems Analysis	0.033	0.030	0.025	22
Relay Systems	0.048	0.049	0.051	6

In the evaluation of Energy R & D technologies, however, the rank order remained as scale-dependent as before, see Table 4.9. Rank preservation of the terminal scores could eventually be guaranteed by the introduction of a uniform scale (γ = 0.50) for the evaluation of the criteria (see Section 4.2). Nevertheless, the terminal scores of Table 4.8 have been used extensively as a basis for discussions in the Council and as a tool for the allocation of an Energy R & D budget. The key issue in our project was to find an allocation which did not (strongly) depend on the scale.

Table 4.8 *Weights of the criteria and AHP terminal scores of the alternative Energy R & D technologies assessed on the basis of their possible contributions to the predominant goals of a national Energy R & D policy. The procedure in the nineteen eighties: the gradations of relative importance and relative performance are simultaneously converted into numerical values on two or three different geometric scales. Aggregation of ratio information via the calculation of arithmetic means.*

	Varying scales for gradations of comparative judgement					
Criteria	$\gamma = 0.72$		$\gamma = 1.00$		$\gamma = 1.44$	
Potential to increase	Weights	Rank	Weights	Rank	Weights	Rank
Security of Energy Supply	0.249	2			0.321	2
Energy Efficiency	0.129	4			0.086	4
Social Acceptability	0.088	6			0.040	6
Innovative Industry	0.187	3			0.181	3
Long-Term Energy Management	0.252	1			0.327	1
High-Level Scientific Activity	0.094	5			0.046	5
Alternatives	$\gamma = 0.72$		$\gamma = 1.00$		$\gamma = 1.44$	
Energy R & D Technologies	Scores	Rank	Scores	Rank	Scores	Rank
Energy Saving in Industry	0.053	4			0.058	4
Energy Saving in Houses, Offices	0.059	2			0.086	2
Energy Saving in Transport	0.037	19			0.026	20
Energy Saving in Other Areas	0.052	5			0.053	6
Oil and Natural Gas	0.043	11			0.044	9
Coal Winning	0.046	8			0.055	5
Coal Combustion	0.044	10			0.047	8
Coal Gassification	0.055	3			0.068	3
Coal Residues	0.039	14			0.034	13
Uranium Thermal Reactor	0.042	12			0.040	12
Uranium Fission Cycle	0.046	7			0.047	7
Uranium Breeder	0.068	1			0.111	1
Uranium Nuclear Fusion	0.037	18			0.029	17
Solar Energy, Thermal	0.037	17			0.029	18
Solar Energy, Electrical	0.045	9			0.042	11
Wind Energy, Decentralized	0.036	20			0.027	19
Wind Energy, Centralized	0.039	15			0.033	15
Biomass Energy	0.038	16			0.029	16
Geothermal Energy	0.030	23			0.018	23
Electricity Production	0.035	21			0.025	21
Thermal Energy Storage	0.040	13			0.034	14
Systems Analysis	0.034	22			0.024	22
Relay Systems	0.048	6			0.044	10

We therefore considered the scale-dependent knapsack problem of maximizing the total benefits

$$\sum_{j=1}^{n} t_j(\gamma) x_j$$

subject to the budget constraint

Table 4.9 *Weights of the criteria and AHP terminal scores of the alternative Energy R & D technologies assessed on the basis of their possible contributions to the predominant goals of a national Energy R & D policy. The procedure in* 1990: *the gradations of relative importance and relative performance are simultaneously converted into numerical values on two or three different geometric scales. Aggregation of ratio information via the calculation of geometric means.*

Criteria	Varying scales for gradations of comparative judgement					
	$\gamma = 0.72$		$\gamma = 1.00$		$\gamma = 1.44$	
Potential to increase	Weights	Rank	Weights	Rank	Weights	Rank
Security of Energy Supply	0.249	2			0.321	2
Energy Efficiency	0.129	4			0.086	4
Social Acceptability	0.088	6			0.040	6
Innovative Industry	0.187	3			0.181	3
Long-Term Energy Management	0.252	1			0.327	1
High-Level Scientific Activity	0.094	5			0.046	5
Alternatives	$\gamma = 0.72$		$\gamma = 1.00$		$\gamma = 1.44$	
Energy R & D Technologies	Scores	Rank	Scores	Rank	Scores	Rank
Energy Saving in Industry	0.054	4			0.060	4
Energy Saving in Houses, Offices	0.058	2			0.079	2
Energy Saving in Transport	0.037	18			0.027	19
Energy Saving in Other Areas	0.051	5			0.050	6
Oil and Natural Gas	0.043	11			0.042	9
Coal Winning	0.045	9			0.054	5
Coal Combustion	0.044	10			0.047	8
Coal Gassification	0.056	3			0.073	3
Coal Residues	0.040	13			0.035	13
Uranium Thermal Reactor	0.042	12			0.041	11
Uranium Fission Cycle	0.047	6			0.048	7
Uranium Breeder	0.067	1			0.114	1
Uranium Nuclear Fusion	0.037	19			0.026	21
Solar Energy, Thermal	0.037	17			0.029	16
Solar Energy, Electrical	0.046	7			0.042	10
Wind Energy, Decentralized	0.036	20			0.028	17
Wind Energy, Centralized	0.040	14			0.035	14
Biomass Energy	0.039	15			0.032	15
Geothermal Energy	0.030	23			0.020	22
Electricity Production	0.036	21			0.027	20
Thermal Energy Storage	0.037	16			0.027	18
Systems Analysis	0.031	22			0.019	23
Relay Systems	0.045	8			0.036	12

$$\sum_{j=1}^{n} p_j x_j \leq B,$$

with zero-one decision variables x_j to indicate whether the j-th technological Energy R & D area is to be supported or not. The coefficients $t_j(\gamma)$ in the objective function were the terminal scores in Table 4.8 representing the benefits of the respective technological areas. The coefficients p_j were the required support levels displayed in Table 4.6, and the right-hand side B stands for the total budget allocated to Energy R & D. With the scale parameter γ varying between 0.72 and 1.44 and the total budget B between Dfl 300 million and Dfl 400 million (a plausible range of values below the required budget of Dfl 438 million), the optimization runs demonstrated again the robustness which is necessary for a workable tool in decision analysis. The technologies which were usually dropped from consideration, regardless of the values of γ and B, had a high cost/benefit ratio: Energy Saving in Industry, Uranium Thermal Reactors, and Uranium Breeders (Légrády et al., 1984). We will present a detailed treatment of the budget-allocation problem in Chapter 7, where we also consider the possible variations in the required support levels.

4.4 SERIOUSNESS OF DISEASES

We present a particular case study here in order to show that the conversion of objective data into subjective grades or scores on a geometric scale is feasible in practice. It was a significant step in a project for public-health planning to assess the seriousness of diseases.

Background information. Since the early eighties attempts are made in The Netherlands to develop a systematic approach towards priority setting in the public-health sector. The points of major concern for the responsible authorities are (a) the identification of criteria to assess the urgency of the emerging public-health problems, and (b) a proper utilization of the epidemiological data in order to set priorities in policies for cure, care, and prevention. In 1993 - 1994, the Ministry of Health, Welfare, and Sport in The Hague commissioned the Delft University of Technology to support a Multi-Criteria Decision Analysis in a Group Decision Room, in order to assess the seriousness of a large number of (clusters of) diseases. Van Gennip et al. (1997) present a more detailed description of the project. The assessment was carried out by twelve high-ranking officials of the Ministry in two rounds, one with the criteria which are of predominant interest for policy makers, and one with the criteria which are highly relevant for patients. The total project, if we do not account for the preparations within the Ministry, took three sessions in the Group Decision Room of the Faculty of Systems Engineering and Policy Analysis. The available tool was GroupSystemV (Ventana Corporation, Phoenix, Arizona) with 15 networked PC's in a U-shape arrangement and a public-screen display, visible for all participants, for electronic brainstorming and voting, see Figure 4.2.

Scope of the first round. Preliminary discussions in the preparatory phase brought up the following criteria to assess the seriousness of diseases from the viewpoint of policy makers in the public-health sector:

1. Projected prevalence (incidence $\times$ average duration) of the diseases in the sample year 2010.

2. Projected number of potential years of life lost in the sample year 2010, as an indicator for mortality.

3. Cost of delivered cure, care, and prevention in the sample year 1988.

Figure 4.2 *Group Decision Room with networked PC's in a U-shape arrangement and a public screen for electronic brainstorming and voting. An appealing feature of the GDR is the possible anonimity of the discussions.*

A major source of information was the policy document "*Public Health Status and Forecasts, the Health Status of the Dutch Population over the Period 1950 - 2010*", published by the National Institute of Public Health and Environmental Protection (1994), and a major issue was how to convert the objective epidemiological data into impact grades expressing the seriousness of the diseases under the above criteria separately. Initially, the criterion weights were taken to be equal since there were no obvious reasons to deviate from equal importance.

For the numerical scaling of the objective performance data the essential information to be supplied by the decision makers is the pair of endpoints of the range of acceptable performance data, under each of the criteria separately. The partitioning of the range between the endpoints into subjectively equal subintervals follows immediately from the choice of the progression factor. Before the sessions in the Group Decision Room the participants could easily obtain an agreement on the endpoints.

The subsequent choice of the progression factor was carried out as follows. Van Gennip presented several possible bars, each of them partitioned in a particular way. Two bars are shown in Figure 4.3.

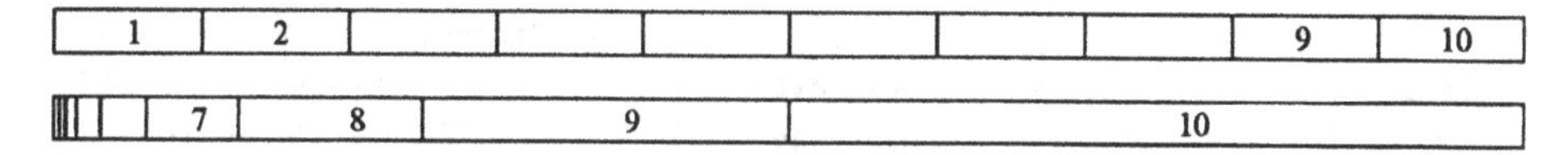

Figure 4.3 *Two possible partitionings of a range representing the context of the seriousness of diseases. The first bar has objectively equal subintervals, the second bar has subintervals demarcated by a geometric sequence of gridpoints with progression factor 2.*

Next, the participants were asked to select a bar, one under each criterion separately. They rejected the bar with objectively equal subintervals because it was felt to be too detailed for the diseases with a high prevalence, mortality, or cost. Eventually, they accepted the bar with the progression factor 2. Table 4.10 shows the partitioning adopted under the respective criteria.

Table 4.10 *Initial subinterval and range representing the context of the seriousness of diseases. Each range is subdivided into* 10 *subintervals. The gridpoints demarcating the subintervals constitute a geometric sequence with progression factor* 2.

Criterion	Initial subinterval	Range
Prevalence in 2010 (inc. × av. dur.)	0 - 3,000	0 - 1,536,000
Years of life lost in 2010	0 - 750	0 - 384,000
Cost of health care in 1988 (Dfl 10^3)	0 - 7,500	0 - 3,840,000

Aggregation in the first round. Given the epidemiological data of the "*Public Health Status and Forecasts*" just mentioned, it was a straightforward step to assign impact grades to the current (clusters of) diseases. Next, the diseases were ranked according to the final grades. The rank order is shown in Table 4.11. Several weight variations were applied thereafter in an elaborate sensitivity analysis. It so happened that a particular collection of fourteen (clusters of) diseases emerged, slightly different from the initial ranking, as the robust set of the most serious diseases from a policy maker's view. For this collection the group of officials judged the effectiveness of prevention, cure, and/or care in a separate session. A description of this step in the evaluation process is beyond the scope of the present volume, however.

The second round. The patient's experience of the burden of a disease was the leading principle in the second round. After some preliminary discussions and experiments the officials eventually decided to rank the diseases under each of the following criteria separately: reduced functioning in social and/or professional life, mental and/or bodily pain, uncontrolled behaviour, reduced memorial capacity, reduced coarse-motorial capacity, and reduced sensorial (perceptual) capacity.

Indeed, ranking seems to be the only appropriate technique here. It is impossible to estimate in real magnitudes how a patient experiences his/her reduced capacities. Ranking is easily supported by GroupSystem V. Each official, working on his/her own PC, moves the diseases in the list upwards or downwards until the desired position is reached. The average rank-order positions eventually produced a group ranking of the diseases. Combining the results of the two rounds, not in a precise manner via a particular choice of criterion weights, but again via an elaborate sensitivity analysis and ample discussions, the group concluded that the chronically invalidating diseases deserved a higher priority. The criteria of the first round did not sufficiently take into account the patient's experience of incurable and disabling diseases.

The last step was also carried out in the Group Decision Room. For the fourteen most serious diseases the officials estimated the effectiveness of prevention, cure, and care (the means of a health policy) under three criteria: prolongation of the healthy life expectancy, improvement of the quality of life for the diseased and the handicapped, and reduction of premature death (the goals of a health policy). A thorough description of this part of the planning process is beyond the scope of the present volume, however.

The results of the project were extensively used in the preparation of a Policy Document submitted to Parliament in May 1995. The Document gratefully acknowledges the importance of the experiments: "*The analysis in the Group Decision Room produced a more precise formulation of the objectives of health policy, and it contributed significantly to priority setting in prevention and research*". In general, such an analysis of the criteria is exactly what MCDA pretends to offer.

Table 4.11 *Impact grades and final grades assigned to the (clusters of) diseases under the criteria of prevalence, mortality, and cost of delivered health care.*

	Prevalence in 2010		Years lost in 2010		Cost care in 1988		final
	$\times\ 10^3$	grade	$\times\ 10^3$	grade	$\times\ 10^3$	grade	grade
coronary heart diseases	354	8	322	10	815	8	8.7
stroke	199	7	138	9	1151	9	8.3
accidents (total)	? high	> 7	130	9	1377	9	> 8.3
chronic obstruct. lung disease	667	9	67	8	450	7	8.0
diabetes mellitus	378	9	50	8	442	7	8.0
dementia	142	7	7	5	1355	9	7.0
lung cancer	24	4	146	9	243	7	6.7
breast cancer	84	6	78	8	187	6	6.7
colon/rectum cancer	52	6	61	8	224	6	6.7
arthrosis	1046	10	low	1	504	8	6.3
pneumonia/ac. bronchitis	38	5	28	7	279	7	6.3
mental handicaps	100	7	low	1	2737	10	6.0
neonatal complications	?	?	50	8	224	6	> 4.7
rheumatoid arthritis	106	7	4	4	159	6	5.7
depression	364	8	low	1	618	8	5.7
Parkinson disease	47	5	11	6	175	6	5.7
hearing disorders	638	9	low	1	176	6	5.3
prostate cancer	22	5	22	6	84	5	5.3
(duodenal) cancer	103	7	5	4	91	5	5.3
schizophrenia	112	7	low	1	587	8	5.3
vision disorders	312	8	low	1	222	6	5.0
stomach cancer	10	3	33	7	87	5	5.0
syndrome of Down	? 13	? 4	7	5	?	?	> 3.0
non-hodgkin lymphoma	9	3	18	6	62	5	4.7
constitutional eczema	397	9	low	1	24	3	4.3
multiple sclerosis	14	4	4	4	93	5	4.3
osteoporosis	50	6	low	1	115	5	4.0
ac.cystitis	29	5	low	1	122	6	4.0
AIDS	0.8	1	7	6	42	4	3.7
oesophagus cancer	0.8	1	13	6	27	3	3.3
cold	19	4	low	1	99	5	3.3
contact dermatitis	97	7	low	1	6	1	3.0
influenza	8	3	3	3	19	3	3.0
sepsis	0.6	1	5	4	39	4	3.0
meningitis (bacterial)	low	1	3	3	37	4	2.7
ac. gastro enteritis	2	1	low	1	44	4	2.0
inflamm. intestinal disorders	low	1	low	1	37	4	2.0

Note on the input facilities. The projects described in the last two sections had a totally different tempo. The participants in the assessment of the Energy R & D areas could leisurely answer the questionnaires of Figure 4.1 in their own office or at home, and return them via the mail. Analysts worked out the encoded judgemental statements in order to prepare a short summary for the next meeting of the Energy Research Council. The decision-making process took many months, the participants had ample opportunities to review their judgement, but there was a limited amount of communication. A session in a

Group Decision Room, however, has more impact (see DeSanctis and Gallupe, 1987; Dennis et al., 1988). The preparations for the assessment of the diseases took also many months, but as soon as the session had been held there was a certain commitment among the participants, possibly due to the intense communication. The judgemental statements could not easily be reviewed thereafter.

4.5 ARITHMETIC AND GEOMETRIC MEANS

In the previous chapters, and also in the present one, we have extensively considered performance ratios and weighted geometric means, concepts which do not immediately appeal to the users of MCDA. Via a number of examples we therefore sketch here the reasons why they have been introduced. We do not have to go deeply into measurement theory (Roberts, 1979) in order to discuss the critical issues.

Let us first recall that, in order to express the position of a point on a one-dimensional axis by a real number, the so-called coordinate, we need two anchor points or, equivalently, one anchor point and a unit of measurement, so that there are in general two degrees of freedom in the coordinates. Ratios of interval lengths on that axis do not depend on the anchor points, however. Usually referred to as dimensionless quantities, ratios of interval lengths are uniquely expressed as ratios of differences of the corresponding coordinates. If one of the anchor points is a natural zero (zero length, zero weight, etc.), there is one degree of freedom only, the unit of measurement, so that ratios of distances from zero, uniquely expressed as ratios of the corresponding coordinates, are also dimensionless quantities. With these considerations in mind we discuss the following examples.

A case where arithmetic means are meaningful. Consider two locations L_j and L_k, and the respective temperatures t_{ij} and t_{ik} at these locations measured at noon on the days $i = 1,\ldots, m$. On some days the observations were made by a European who recorded the temperatures in degrees Celsius, on the remaining days by an American who noted the temperatures in degrees Fahrenheit. The average difference of the noon temperatures at these locations, given by the arithmetic mean

$$\frac{1}{m}\sum_{i=1}^{m}\left(t_{ij}-t_{ik}\right),$$

is a quantity which makes sense for a decision maker if the observations are rescaled so that all temperatures are expressed in the same units of measurement.

A case where geometric means are meaningful. Let us consider now two programmed algorithms A_j and A_k, and the respective execution times t_{ij} and t_{ik} to solve the test problems $i = 1,\ldots, m$. Some of these execution times are expressed in seconds (for the small test problems), others in minutes or hours (for the large problems). Even after a conversion into the same time units, it is meaningless for a decision maker to compare the efficiency of the algorithms on the basis of the arithmetic mean of the $t_{ij} - t_{ik}$, because the execution times of the large test problems will be dominant. This is also true for the ratio of the arithmetic means (or the sums) of the execution times which is given by

$$\frac{\sum_{i=1}^{m} t_{ij}}{\sum_{i=1}^{m} t_{ik}}.$$

This ratio does not properly show how much slower A_j is with respect to A_k because the large test problems will be decisive. The geometric mean of the execution times

$$\sqrt[m]{\prod_{i=1}^{m} \left(\frac{t_{ij}}{t_{ik}}\right)}$$

makes sense, however, at least under the additional assumption that all test problems have the same importance because each of them represents an unknown number of practical problems. The inverse expresses globally the relative efficiency of the programmed algorithms. The numerical value of the geometric mean does not depend on the time units since all ratios in it are dimensionless quantities (time has a natural zero here).

Geometric means on a new, common dimension. We consider again two programmed algorithms A_j and A_k, the respective execution times t_{ij} and t_{ik} to solve the test problems $i = 1, \ldots, m$, as well as the associated space requirements s_{ij} and s_{ik}, some of them expressed in kilobytes, others in megabytes. The geometric mean

$$\sqrt[2m]{\prod_{i=1}^{m} \left(\frac{t_{ij}}{t_{ik}} \times \frac{s_{ij}}{s_{ik}}\right)}$$

does not immediately make sense for a decision maker, although the ratios in the above expression are dimensionless quantities. Some of them are ratios on the dimension of time (seconds, minutes, hours), others are ratios on the dimension of space (kilobytes, megabytes). The geometric mean makes sense, however, if we make a transition to a new, common dimension such as quality or desirability. If "p times faster" means "p times more desirable", for instance, if ratios of space requirements are interpreted in the same way, and if all ratios have the same importance, then the inverse of the above geometric mean may be used to stand for the desirability ratio of the programmed algorithms.

Geometric means and compatible reference points. Suppose that a decision maker considers the locations L_j and L_k again and that he/she asks how much warmer L_j is with respect to L_k. Ratios of recorded temperatures do not make sense unless we use the Kelvin scale, but ratios of differences with respect to reference points on the Celsius or Fahrenheit scale can provide meaningful information. The reference points should be compatible, however. The freezing points 0°C and 32°F of water constitute such a pair of reference points, for instance, as well as the lower limits –18°C and 0°F where outdoors human life begins to be feasible until the upper limits 38°C and 100°F are attained. Taking the symbol $t_i^{\min}$ to represent the reference point at the i-th day (18°C or 64°F if we consider two summer-holiday locations, for instance, and –18°C or 0°F for two winter-sport locations), we can reasonably use the geometric mean

$$\sqrt[m]{\prod_{i=1}^{m}\left(\frac{t_{ij}-t_i^{\min}}{t_{ik}-t_i^{\min}}\right)}$$

to express the relative warmth of L_j with respect to L_k, at least if the reference points are below the recorded temperatures. We can also assign normalized weights c_i to the respective days in order to express the relative importance of certain peak periods. The relative climatic attractiveness of the two locations will accordingly be given by

$$\prod_{i=1}^{m}\left(\frac{t_{ij}-t_i^{\min}}{t_{ik}-t_i^{\min}}\right)^{c_i}.$$

Note that we do not have to rescale the recorded temperatures. If they are all expressed in the same units, however, we only have to specify one reference point $t^{\min}$.

With these considerations in mind, we introduced the abstract dimension of desirability as the new, common dimension for an aggregation procedure, see Section 2.5. Taking D_{ij} and D_{ik} to represent the desirability of the alternatives A_j and A_k under criterion C_i, and the maximum desirability $D^{\max}$ as the reference point, we modelled the global preference for A_j with respect to A_k as the weighted geometric mean

$$\prod_{i=1}^{m}\left(\frac{D^{\max}-D_{ik}}{D^{\max}-D_{ij}}\right)^{c_i}$$

with normalized criterion weights c_i. This is in fact how we aggregate, regardless of whether we have quantitative or qualitative criteria. Obviously, we use the inverse ratios of the deviations from the maximum desirability.

REFERENCES TO CHAPTER 4

1. Belton, V., and Gear, A.E., “On a Shortcoming of Saaty’s Method of Analytical Hierarchies”. *Omega* 11, 227 – 230, 1983.

2. Dennis, A., George, R.J.F., Jessup, L.M., Nunamaker, J.F., and Vogel, D.R., "Information Technology to Support Electronic Meetings". *MIS Quarterly* 12, 591 - 624, 1988.

3. DeSanctis, G., and Gallupe, R.B., "A Foundation for the Study of Group Decision Support Systems". *Management Science* 33, 598 – 609, 1987.

4. Gennip, C.G.E., Hulshof, J.A.M., and Lootsma, F.A., "A Multi-Criteria Evaluation of Diseases in a Study for Public-Health Planning". *European Journal of Operational Research* 99, 236 – 240, 1997.

5. Légrády, K., Lootsma, F.A., Meisner, J., and Schellemans, F., "Multi-Criteria Decision Analysis to Aid Budget Allocation". In M. Grauer and A.P. Wierzbicki (eds.), *"Interactive Decision Analysis"*. Springer, Berlin, 1984, pp. 164 – 174.

6. Lootsma, F.A., "Modélisation du Jugement Humain dans l'Analyse Multicritère au Moyen de Comparaisons par Paires". *RAIRO/Recherche Opérationnelle* 21, 241 - 257, 1987.

7. Lootsma, F.A., "Scale Sensitivity in the Multiplicative AHP and SMART". *Journal of Multi-Criteria Decision Analysis* 2, 87 – 110, 1993.

8. Lootsma, F.A., Meisner, J., and Schellemans, F., "Multi-Criteria Decision Analysis as an Aid to the Strategic Planning of Energy R & D". *European Journal of Operational Research* 25, 216 – 234, 1986.

9. Roberts, F.S., *"Measurement Theory"*. Addison-Wesley, Reading, Mass., 1979.

CHAPTER 5

THE ALTERNATIVES IN PERSPECTIVE

The question of what preference actually is and how it is triggered has rarely been discussed in the literature on MCDA although it is of prime importance. How do we see the alternatives? Methods for MCDA are designed to reveal the (pre-existing?) preferences of the decision makers via more or less sophisticated elicitation procedures. Since the true nature of preference is not a point of discussion there are several schools in MCDA. Some of them advocate ordinal methods with the argument that the decision makers cannot accurately measure their preference intensities. Others, however, develop cardinal methods because they are convinced that subjective measurement of preference intensity is feasible and sufficiently reliable indeed. Thus, the unresolved issue of what preference actually is and how it can be modelled strongly affects the development of MCDA methods. For an extensive discussion of the issue we refer the reader to the Chapters 7 and 10 of Von Winterfeldt and Edwards (1986).

5.1 A STATIC PERSPECTIVE

Let us first summarize the categorization of a range of values introduced in Chapter 2. In order to model the preferences of the decision maker we assume that decisions are invariably made within a particular context. Since quantitative criteria are usually one-dimensional, in the sense that the performance of the alternatives can be expressed in physical or monetary units on a one-dimensional scale, we model the context as the range of acceptable performance data on the corresponding dimension. The lower and the upper bound are to be estimated by the decision maker. Next, starting from a particular viewpoint we partition each of these ranges into a number of subintervals which are subjectively equal. On the basis of psycho-physical arguments the gridpoints demarcating the subintervals constitute a geometric sequence. The behaviour of the gridpoints is in fact surprisingly uniform. The progression of historical periods and planning horizons, the categorization of nations on the basis of the population size, and the categorization of light and sound intensities, they all have a progression factor which is roughly equal to 2. This enables us to categorize the ranges and to quantify verbal expressions like weak, strict, and strong preference, the so-called gradations of relative preference. Thus, the leading idea of the approach is that the decision maker, standing at the so-called desired target, views the range of acceptable performance data in perspective.

Qualitative criteria, however, are mostly low-dimensional in the sense that the performance of the alternatives can only be expressed in several more or less verbal terms. The pleasure of driving a car, for example, depends on the maximum speed, the maximum acceleration, and the level of noise and vibrations. The aesthetic attractiveness of the car body design depends on the decision maker's varying viewpoint. In such cases, one can ask the decision makers to rate the alternatives directly. First, they have to determine the endpoints of an interval on a one-dimensional scale. They are requested to identify the worst and the best (real or imaginary) alternative and to assign proper grades to them. Thereafter they may interpolate the remaining alternatives between the endpoints. They may alternatively introduce a numerical scale (a seven-point scale, for instance) and they may verbally describe the significance of the echelons. This is not always an easy procedure but since it is current practice in schools and universities we assume that it is workable, at least for experienced decision makers.

In Figure 2.4 we can immediately see that the proposed categorization under quantitative criteria is in fact a very simple operation. In the horizontal direction one starts with a small initial step which is repeatedly doubled until the opposite end point is reached. An alternative way of looking at the operation is to consider it as a series of bisections of the range. In the vertical direction one just counts the number of steps. One does not need a particular unit of measurement to carry out these operations. Bisections and duplications, the progression factor 2, they will certainly remind the reader of the partitioning of time in musical compositions: tone durations as bisections or duplications of the rhythm's fundamental beat and repetitions of phrases, possibly transposed, in the melody. Bisections and duplications seem to be so deeply embedded in human behaviour that they may have a physiological basis in the neural system. There is a certain rhythm in the progression of judgemental gradations, a rhythmic pattern which is also found in the subjective measurement of time, as we will presently see.

The categorization of Chapter 2 can also be described as follows. The length of the range is set to 64 units, and the major grid points are situated at 4 and 16 units from the beginning of the range of acceptable performance data. Such a categorization can be observed in the landscape around us, as the straight railway track of Figure 5.1 readily shows. The 4 km distance with respect to the horizon, demarcated by the high-voltage line carriers which are 70 - 75 meters apart, has been subdivided into roughly 60 units. There is a foreground zone until the carrier 4, a middle zone between the carriers 4 and 16, and a background zone thereafter.

Figure 5.1 *Subdivision of a range of roughly* 60 *units of measurement. There is a foreground zone until the carrier* 4, *a middle zone between the carriers* 4 *and* 16, *and a background zone thereafter.*

It is well-known that the perspective in a painting has to be carefully constructed. Wadum (1995), for instance, described how the painter Johannes Vermeer (1632 - 1675) prepared the "Lady at the Virginals with a Gentleman" via the choice of vanishing points and the drawing of certain diagonal lines. Figure 5.2 shows the framework which enabled Vermeer to approximate the reality with an almost photographic accuracy. Vermeer pressed a pin in the canvas, at the central vanishing point where certain diagonal lines meet (the lines parallel to the line of sight). Thereafter he connected a chalk-powdered string to the pin, drew it up, and let it snap back so that it left a straight line of chalk on the canvas. Figure 5.3 makes it clear that the harmonious interior has not been loosely painted. The harmony is the result of the accurate construction. This also encouraged us to support preference modelling by the careful construction of geometric scales within the ranges of acceptable performance data representing the context of the actual decision problem.

In essence, the measurement of preference intensities is closely related to the subjective measurement of space and time, to the perception of the environment wherein we live as a world in motion. Indeed, when we express our preferences we measure or estimate the performance of the alternatives, both at present and in future circumstances. We estimate (sometimes in a fraction of a second) what the alternatives will afford us to do and/or to enjoy. In fact, we have powerful tools to do so because subjective measurement is an integral part of our daily life. We continually measure the past, present, and future position of the objects around us with respect to our body. We continually perceive the environment and ourselves in it. With some additional professional guidance the quality of subjective measurement can be astonishing: see how a tennis player coordinates his/her observations and actions, and hear how the members of an orchestra synchronize their movements following the rhythm of the musical composition.

The environment is more or less tangible and/or visible but time is volatile. Living creatures have a surprising ability, however, to control a time-dependent series of actions. Rhythmic activities like walking, running, and tapping are controlled by a timekeeper which has a hierarchical organization. It is difficult for human beings, for instance, to tap different rhythmic patterns with the two hands unless one of the two patterns is an integral multiple of the other (Peper, 1995). The timekeeper is also amodal, in the sense that it extends across communication channels. Tapping a rhythmic sequence with one hand while listening to a sequence of tones with a different rhythmic structure leads to disruptions in performance. In summary, time can subjectively and physically be measured by rhythmic movements. The annual rotation of the earth around the sun and the daily rotation of the earth around its axis set the pattern of human life, whereas the pendulum further partitions the days into hours, minutes, and seconds.

Time can also be measured by the rate of change of certain processes. Many living creatures have a biological or physiological clock to measure the time which elapsed since a particular moment. Bünning (1973) presented several examples of actions and processes which are triggered, not by the transition from the night to the day, but by the length of the day. Human beings have a circadian rhythm of approximately (not exactly) 24 hours so that there is a daily phase shift between the physiological rhythm and the rotation of the earth (Winfree, 1987). The physiological clock, which is usually overruled by the 24 hours period, resumes control as soon as the person in question is isolated from the rhythm of light and dark.

Time and preference are volatile but if time can subjectively be measured the gradations of preference may be measurable as well, at least as soon as an appropriate viewpoint has been found. From there the spectator or the decision maker will have a static perspective on the alternatives.

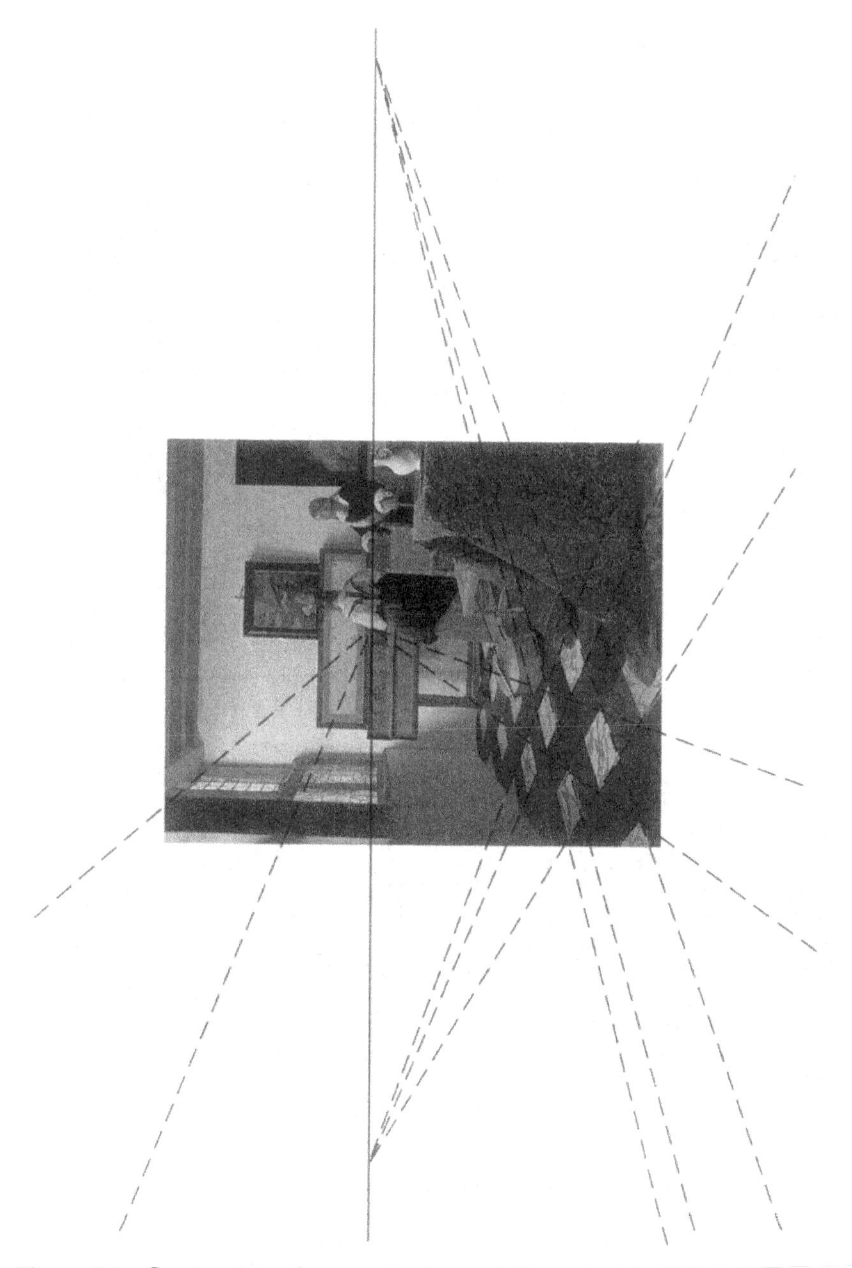

Figure 5.2 *Construction of perspective by vanishing points and diagonal lines in the "Lady at the Virginals with a Gentleman" by Johannes Vermeer (1632 - 1675).*

Figure 5.3 *The "Lady at the Virginals with a Gentleman" by Johannes Vermeer. The harmonious interior with an almost photographic accuracy as the result of the careful construction of the perspective.*

5.2 VISUAL PERCEPTION AND MOTOR SKILLS

The static perspective is inadequate, however. Human beings are dynamic spectators. They perceive in order to move, but they also move in order to perceive (Gibson, 1979; Kelso, 1982; Rosenbaum, 1991). This is due to the visual effects of retinal stabilization. Usually, when an eye moves, the image of a stationary object shifts across the back of the eye, the so-called retina. Images of moving objects also undergo some retinal slippage since visual tracking tends to be imperfect. When retinal slippage is eliminated, however, subjects see parts of the image disappear. Ultimately, the image vanishes completely. The light-sensitive cells in the retina fatigue or adapt rapidly if they are continually stimulated in an unchanging fashion. The motion of the eye prevents this fatigue or adaptation. This experiment can be carried out by every human being. One just has to keep the eyes in a fixed position and one only has to concentrate on a particular object in order to see that the image gradually disappears. This is a laboratory experiment, however. The eyes do not in reality fixate, but they undergo many saccadic movements, rapid jumps from one position to another where they are in a delicate state of equlibrium for a fraction of a second.

The physiological necessity to move goes hand in hand with the necessity to maintain a panoramic awareness of the environment. This can more easily be explained if we compare the processes of acoustic and visual perception. Audible sounds carrying information about their sources and about the surfaces where they have been reflected come from all directions. They turn around corners, penetrate through walls, and even if one does not move, one can get a spherical awareness of the environment. Visible light, however, travels in straight lines, it only penetrates through glass walls, and the field of vision of human beings is slightly smaller than a hemisphere. Human beings therefore have to move in order to see the environment panoramically. In the ecological theory of visual perception (Gibson,1979; see also the critical discussion in Bruce and Green, 1990) human beings perceive the environment, not with their eyes only, but with a visual system consisting of the eyes in the head on the body on the ground. With such a system human beings become aware of the surrounding environment and of themselves in it. In order to perceive an object, they approach it, they walk around it, and they observe it under different angles. The impulse to do so is irresistible. They move in order to perceive the invariants in the changing perspective: the horizon, the occluding edges separating the hidden and the unhidden surfaces, the reflectance and the texture of the unhidden surfaces, the connections of the straight and the curved lines, and certain ratios. Note, for instance, that the horizon cuts the carriers on the right-hand side of the railway track in Figure 5.1 all in the same ratio depending on the height of the observer.

The dynamic perception of the alternatives and the search for a proper perspective may easily lead to incomparability, intransitivity, and cyclic judgement of the alternatives under a qualitative criterion. This can more easily be explained by the visual arts than by a thousand words. Consider the Waterfall (1961), a well-known lithograph of Maurits C. Escher (1898 - 1972), reproduced in Figure 5.4. The central role is played by a water stream. It drives a water wheel and then it follows a slightly downward direction. Snaking through the towers it returns to the spot from where it falls again to drive the wheel. By following the continually changing perspective (downstream, then upwards, the urge is irresistible) the spectator perceives the point furthest away as the one which is also closest and the highest point as the one which is also lowest. In this way the waterfall achieves what we regard as impossible: a *perpetuum mobile*. We recognize this as soon as we holistically perceive the twin towers and the waterfall from a certain distance. In the dynamic pursuit of the water flow, however, our perspective is too narrow. Thus, intransitivity and cyclic judgement may occur because human beings are dynamic spectators, in search of a proper perspective. Most decision makers will have the experience that it may be difficult to find an appropriate standpoint in matters with important ethical and moral aspects. The Waterfall suggests that they may be trapped in a cycle before they eventually find a viewpoint where they can draw their conclusions.

The assumption that the gradations of preference are shaped by the decision maker's dynamic perspective on the range of acceptable performance data has important implications for decision analysis. First, in their daily life human beings have to learn how to use the perspective in order to develop their motor skills. Hence, they also need professional training in order to decide whether they have chosen the relevant criteria, the relevant viewpoints, and the relevant ranges of acceptable performance data. Second, preferences are not completely fixed. Although human beings may be able to measure their preferences from a certain viewpoint with a high accuracy, they continually change their viewpoint in order to identify the invariants in their preferences. It may even happen that some alternatives are incomparable because the viewpoints which seem to be relevant today and those that may be relevant in one or more future scenarios lead to conflicting conclusions. Third, there is an important implication for group decision making. In order to achieve agreement with other members, in order to share their preferences and their subjective measurement, the group members have to see the problem from the perspective of their colleagues, not in an abstract sense but literally, over the range from their standpoint to their horizon. Hence, they usually have to travel to the hot spot where the actual decision problem emerged.

A striking idea in the ecological theory of Gibson (1979) is that human beings, looking at the surfaces and objects around them, do not perceive the qualities or attributes but the so-called affordances. Do these surfaces provide support for walking and running? Do these objects provide shelter and protection, are they eatable, are they usable for our needs to lift and to carry or to throw things, to write, to draw,...? Indeed, human beings seem to have an open eye for possibilities. The Romance languages have an astonishing amount of words ending on -able or -ible to express what certain objects afford us to do: fruit may be eatable, roads may be passable, porcelain may be breakable, etc. The Germanic languages have similar words ending on -bar or -baar to express what the objects in the environment enable human beings to do.

Figure 5.4 *The Waterfall* (1961), *a lithograph by M.C. Escher* (1898 - 1972). *The spectator who follows the perspective along the stream is surprised by the absurdity of the construction. Each succesive step along the stream seems to be reasonable, but the cycle is impossible.*

Perception is also called a simulated action (Berthoz, 1997). As soon as human beings observe a desirable object, they internally carry out the movements which are necessary to use it. If the affordances are the predominant features of the surfaces and the objects in the environment (note that the ecological theory is still controversial, see Bruce and Green, 1990), then we must completely revise the formulation of the criteria in MCDA, a step which is traditionally based upon the qualities of the alternatives. Such a revision is not necessarily surprising. What do we see in a car, for instance? Qualities such as the sticker price, the engine capacity, the cargo volume,...., or the life-style which it affords? Do we internally carry out the movements to sit down in a desirable car and to drive it? Do the affordances lead us to the real motives of the decision makers?

5.3 COLOUR PERCEPTION

In our approach we basically have to answer the following question: is it reasonable to assume that cognitive processes such as the articulation of preference intensities are shaped by the physiological properties of the perceptual and the motorial system of human beings? In general such a question is difficult to answer. Even the physiological basis of the relations between perception and movement has only recently been studied (Berthoz, 1997). However, there is a particular example which strongly supports the assumption. The example is given by the properties of the system for colour perception.

An interesting result of anthropological research (Berlin and Kay, 1969; Kay and McDaniel, 1976) is that all human languages share a universal system of eleven basic colour terms (black, white, red, green, yellow, blue, brown, grey, orange, purple, pink). During the history of a language the basic colour terms become encoded in a partially fixed order. In other words, the development of a language seems to pass through a number of stages where colour terms are successively added to the available set according to a number of strict rules. Experimentally, the research began with the elicitation of the basic colour terms in the language of the informant representing a particular tribe or people. He/she was asked (a) to select the colour chips which were the best examples of the respective colour terms in his/her language and (b) to indicate the boundaries of each colour category by selecting all chips whose colour was somehow covered by the given term. The first task appeared to be much easier than the second one, and it confirmed that colour categorization is not arbitrary.

The universality of colour categorization has its basis in the physiological system of human beings. Colour perception begins at the retina with the stimulation of the colour-sensitive cones and the black-and-white-sensitive bars. In the neural pathway from the eyes to the brain there are two types of so-called opponent-response cells for colour transmission: the RG cells generate a unique red-or-green response to the stimulus coming from a particular wavelength, the YB cells a unique yellow-or-blue response. Hence, any colour appears subjectively either as a pure instance of one of the four primary colours red, green, yellow, and blue, or as a composite: all colours except the primary ones could be seen to consist of the simultaneous perception of two primaries. Purple is composed of red and blue, chartreuse is a composite of yellow and green, turquoise a composite of blue and green, etc. There is no such thing as the simultaneous perception of green and red, however, nor of blue and yellow. Perceptually, there is no yellowish-blue, for instance.

Colour categorization plays an important role in the area of linguistics (Taylor, 1995) where one studies how human beings translate meaning into sound through the formation of categories, that is, through the vague discretization of the continuum of stimuli coming from the outside world. Categorization is a fundamental activity whereby human beings structure and store their knowledge. The key questions are: do categories have any basis in the outside reality, does the outside world and/or the perceptual system of human beings shape the categories, or are the categories merely arbitrary constructs of the human mind? Since the early decades of this century it was commonly assumed that the categories which human beings perceive in reality are not objectively there. Encoded in the native language, categories have been enforced upon human beings as a product of the learning process in the formative years. This view was supported by the fact that languages differ considerably, both in the number of colour terms they possess and in the range of colours covered by the respective terms. Languages obviously discretize the three-dimensional colour continuum (with hue, brightness, and saturation as its dimensions) in an arbitrary manner. According to the structuralist point of view in linguistics, the arbitrariness of colour categorization is paradigmatic for the arbitrariness of language as a whole. People use language to talk about the outside world, but language is an autonomous, self-contained system with terms which derive their meaning from the simultaneous presence of other terms. Colour perception was frequently used to illustrate this point of view: when a language acquires a new colour term, the whole system changes and the meaning of the previously available terms undergoes a shift. All colour terms in such a system have equal status. Some colour terms might be used more frequently than others, but there are no privileged terms.

The studies of Berlin and Kay (1969) and Kay and McDaniel (1976) presented a serious challenge to the structuralist assumption that the colour spectrum is arbitrarily discretized, from language to language. On the contrary, there is a uniform collection of privileged basic colour terms which are increasingly incorporated in a language during its history. Although the range of colours which is covered by the term *red* may even vary from person to person, for instance, there is a remarkable unanimity on what constitutes a good example of *red*. Thus, colour categorization is conditioned by the physiological properties of the perceptual system. A basic colour term like *red* denotes, first, a focal colour, and through generalization it acquires its full denotational range. The addition of a new term such as *orange* causes the denotational range of *red* to contract but the centre of the category remains unchanged. The arguments brought forward suggest that other cognitive processes such as the articulation of preferences may also be shaped by the physiological properties of the perceptual system.

2.6 AN AESTHETIC PLEASURE?

Preference is not only shaped by the decision maker's dynamic perspective on the alternatives. The geometric sequence of preference intensities, the layout of repeated bisections or duplications, they may also be an aesthetic pleasure to observe. The idea is due to Lipovetsky (see Lipovetsky and the present author, 1999) who explained it via a generalization of the golden section. He partitioned a given line segment into a finite sequence of diminishing subsegments in such a way that the ratio of any pair of adjacent subsegments is equal to the ratio of the whole and the first subsegment. When the number of subsegments increases one approximates the subdivision of the line segment via repeated bisections. The golden section plays an important role in the analysis of beauty and harmony. It appeared in the design of many ancient and modern buildings. Hence, we assume that the categorization of the preference intensities may also yield the experience of aesthetic pleasure.

The golden section partitions a line segment with length L into two subsegments, the major with length M and the minor with length m, such that

$$\frac{L}{M} = \frac{M}{m}.$$

Thus, M is the geometric mean of L and m. In the words of Euclides (4-th century BC), the area Lm of the rectangle spanned by the whole and the minor equals the area M^2 of the square spanned by the major. It is customary to designate the ratio of the whole and the major by the symbol ϕ, the initial letter of the name of the Greek sculptor Phidias (5-th century BC). The term "golden section" or "divine proportion" dates from the early 16-th century, however. Rewriting the above equation in the form

$$\frac{M+m}{M} = \frac{M}{m} = \phi$$

it follows easily that

$$\phi^2 - \phi - 1 = 0,$$

an equation which has a positive and a negative root. The Phidias number ϕ must be positive so that

$$\phi = \frac{1+\sqrt{5}}{2} \approx 1.618.$$

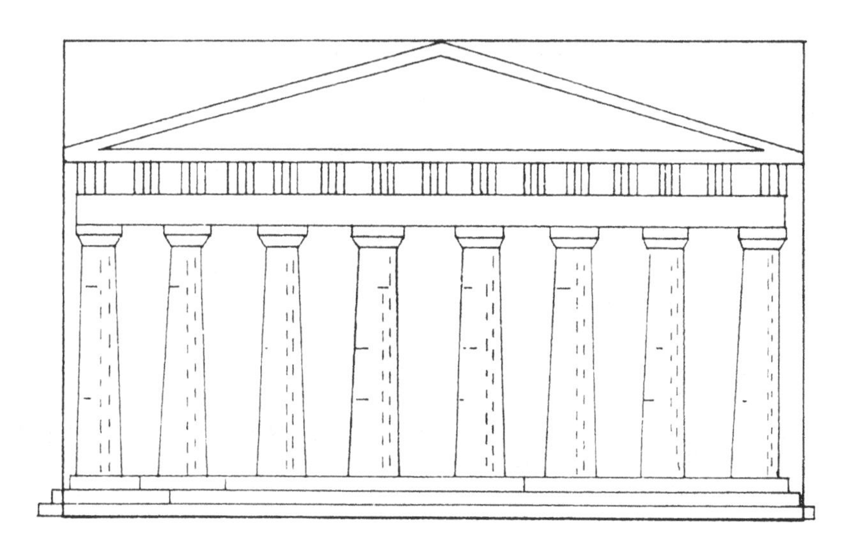

Figure 5.5 *The Parthenon in Athens (fifth century BC). Its dimensions could be fitted almost exactly into a golden rectangle. It therefore stands as another example of the aesthetic value of this particular shape.*

From ancient times the golden section has played an important role in architecture and in the visual arts, particularly in the analysis of aesthetic pleasure. The golden section and the golden rectangle, a figure which is spanned by two sides in the golden section ratio, have been found in the harmonious proportions of temples, churches, statues, paintings, pictures, and fractals (Doczi, 1981; Hagenmaier, 1984; Herz-Fischler, 1987). The front of the Parthenon in Athens (5-th century BC), for instance, fits almost exactly in a golden rectangle. This is shown by Figure 5.5. The golden section was very popular in the period of the Renaissance (15-th and 16-th century), and recently Le Corbusier (1887 - 1965) used it in many architectural designs.

It is not always clear whether the artists consciously used the golden section or whether they intuitively approximated the so-called divine ratio (Beutelspacher and Petri, 1995). Pure symmetry tends to be boring. An asymmetric arrangement of the important components, however, arouses the spectator's interest. It frequently happens that the arrangement is more or less in accordance with the golden section. Several experiments in the judgement of art, when rectangles of varying proportions are submitted to experienced and inexperienced subjects, show that the golden rectangle seems to be particularly attractive (Hekkert, 1995). Finally, note that the golden section has also been found in the structure of musical compositions, in the ratios of harmonious sound frequencies, in the dimensions of the human body, and in the structure of plants (Doczi, 1981; Hagenmaier, 1984).

Let us now consider the subdivision of a line segment with length L into n subsegments $m_0 > m_1 > \ldots\ldots > m_{n-1}$, such that

$$\frac{L}{m_0} = \frac{m_0}{m_1} = \cdots\cdot = \frac{m_{n-2}}{m_{n-1}} \equiv \phi_n \cdot$$

To our knowledge, this generalization is new. In what follows we shall refer to ϕ_n as the progression factor. Obviously, ϕ_2 equals the Phidias number ϕ. Introducing the regression factor or discount factor $\rho_n = 1/\phi_n$ we can easily obtain

$$L = m_0 + m_1 + m_2 + \cdots\cdot + m_{n-1} = m_0(1 + \rho_n + \rho_n^{\ 2} + \cdots + \rho_n^{\ n-1}).$$

In financial planning the right-hand side is typically the present value of n consecutive annual payments of the amount m_0 at the interest rate i_n which has a value such that the discount factor is given by

$$\rho_n = \frac{1}{1 + i_n/100}.$$

However, the discount factor cannot arbitrarily be chosen here because

$$1 + \rho_n + \rho_n^{\ 2} + \cdots\cdot + \rho_n^{\ n-1} = \frac{L}{m_0} = \phi_n = \frac{1}{\rho_n}, \tag{5.1}$$

which implies that ρ_n satisfies the equation

$$\rho^n + \rho^{n-1} + \cdots + \rho - 1 = 0. \tag{5.2}$$

This equation has one change of sign so that there is exactly one positive root for each n (Descartes' rule). Summing up the terms on the left-hand side of (5.1) we obtain

$$\frac{1-\rho_n^{\,n}}{1-\rho_n} = \frac{1}{\rho_n}. \tag{5.3}$$

In the limiting case, when $n \to \infty$, it follows easily from (5.3) that the sequence $\{\rho_n\}$ of regression factors converges to $^1/2$. This limiting value is rapidly attained. The positive root of (5.2), obtained by Cardano's rule for $n = 3$ and via Newton's method for higher values of n, yields the results displayed in Table 5.1. The length L of the initial segment has been set to 1 so that the entries in the respective columns exhibit the proportional subdivision of the initial segment. The last row shows the corresponding values of the regression or discount factors ρ_n and the progression factors ϕ_n. Obviously, from $n = 7$ the generalized golden section practically coincides with a subdivision via repeated bisections of the initial segment.

The perception of time and space, light and sound is fundamental for human life. It provides the basic conditions for the motor skills whereby human beings coordinate and control their movements. We assume that preference intensities have the same pattern of repeated bisections or duplications as the perception of time and space, light and sound. The relationship between the generalized golden section and the subdivision via repeated bisections suggests that it may be an aesthetic pleasure to observe such a pattern.

Table 5.1 *Subdivision of a segment with length* 1 *according to the generalized golden section rule for increasing values of n, the number of subsegments. The regression factors* ρ_n *converge to 0.5, the progression factors* ϕ_n *start from the Phidias number* ϕ *and converge to* 2.

	$n=2$	$n=3$	$n=4$	$n=5$	$n=6$	$n=7$
m_0	0.618	0.544	0.519	0.509	0.504	0.502
m_1	0.382	0.295	0.269	0.259	0.254	0.252
m_2		0.161	0.140	0.132	0.128	0.127
m_3			0.072	0.067	0.064	0.064
m_4				0.034	0.033	0.032
m_5					0.016	0.016
m_6						0.008
ρ_n	0.618	0.544	0.519	0.509	0.504	0.502
ϕ_n	1.618	1.839	1.928	1.966	1.984	1.992

5.4 A COHERENT SET OF CRITERIA

So much has been said about perspectives and viewpoints that we now have to consider the choice of the criteria in a decision problem. The examples in the Chapters 2 and 3 may have given some idea of the requirements to be satisfied by the criteria. We also sketched the difficulties which arise as soon as the decision maker tries to comply with the requirements. Briefly speaking, the analysis should be based upon a coherent and adequate level of detail, not too refined, not too coarse. The criteria should be operational and they should be independent, the number of criteria should be large enough to cover all aspects of the decision problem and it should be small enough to keep the analysis manageable. These are, more or less, the desirable properties listed by Keeney and Raiffa (1976), and Roy (1985) refers to a set of criteria with such properties as a coherent family. Let us discuss the requirements here in detail.

1. The criteria should be operational in the sense that the decision maker must be able to judge the relative performance of the alternatives under each of them. The units of measurement must be meaningful to the decision maker, and he/she must be able to specify the context, as a range on the corresponding dimension, or at least in verbal terms describing the best and the worst alternatives which are in principle acceptable. The endpoints of the ranges must be compatible, as we have seen in the car-selection example of Section 2.6. The decision maker may then express his/her judgement in various ways: in verbal terms (pairwise comparisons), in grades on a numerical scale (direct rating), or in rank order positions. It is not necessary to follow the same procedure throughout the analysis. He/she may choose a cardinal method (pairwise comparisons or direct rating) under the criteria where gradations of preference can be distinguished, and otherwise he/she may turn to an ordinal method (ranking). This implies, of course, that a decision support system for MCDA must have the character of a poly-algorithm which enables the decision maker to use a variety of algorithmic steps. The intermediate results must be aggregated into global information which enables the decision maker to identify the preferred alternative, to classify the alternatives into a small number of categories, or to rank them in a subjective order of preference. Note that intransitivities and cyclic judgement are not necessarily irrational. The criteria where these phenomena occur may still be operational.

2. Independence of the criteria is the tacit assumption underlying the subjective evaluation of the alternatives under a given criterion. Direct rating, pairwise comparisons, and ranking should not depend on the performance of the alternatives under the remaining criteria. Moreover, overlaps between criteria should be prevented as much as possible so that one avoids the double counting of the performance of the alternatives. This requirement is easier to meet when the number of criteria is small. One does not always have to introduce separate criteria for different types of cost or profit, for instance. On many occasions it is sufficient to amalgamate the expenditures for capital investment, maintenance, and operations into one single item: the annual

expenditures. Overlaps cannot always be avoided, however. This has been shown in the staff-appraisal problem of Section 3.7.

3. The set of criteria must be complete. It has to cover all relevant aspects of the decision problem so that the final decision can be made without recourse to additional arguments. An incomplete set of criteria is frustrating for the cooperation between the decision maker and the analyst, but it is not rare. Sometimes the decision maker cannot or does not want to reveal the delicate arguments, and although the analyst feels that there is a hidden agenda he/she may be unable to put the real arguments on the table for discussion. Note that a criterion which is generally thought to be relevant or important may be ineffective in a particular decision problem because it does not discriminate between the alternatives. Hence, it can be dropped. By the introduction of new alternatives with striking properties it may also happen that new criteria become effective. Because the criterion weights are normalized, they will always be affected by the deletion and the addition of criteria. In our projects, however, we assumed that the weight ratios of the other criteria remained unchanged, unless the decision makers requested a new evaluation of the criteria.

4. The set of criteria must be manageable. In essence, we are confronted here with Miller's law (1956) stating that human beings cannot survey more than roughly seven items. The step which is usually recommended to overcome this limitation, the construction of a hierarchy of criteria, sub-criteria, etc., has several pitfalls as we demonstrated in Section 3.7: the hierarchy is not necessarily unique, and the structure considerably affects the final grades and the terminal scores. Hence, we suggest the decision makers to use a small number of criteria only. They may be one-dimensional or low-dimensional, but if they are small in number they can holistically be considered, they can be discussed at length, and the decision maker can really weigh their relative importance. He/she may compare the criteria in pairs, considering the emotional values of the criteria or wondering under which of the two he/she would prefer to improve the percentual (!) performance of the alternatives. He/she may also use the weights which were previously employed on similar occasions so that the consequences are roughly known. This will enhance the consistency of his/her policy. In fact, the decision maker has to work on each criterion. He/she has to establish its relative importance within the total set of criteria, and he/she has to check whether the alternatives can meaningfully be judged from its corresponding viewpoint.

Distributed decision making is frequently found in large organisations. In such a process the criteria and the alternatives are not always assessed by the same decision makers. The evaluation of the alternatives may have been entrusted to a group of technical experts but the formulation and the weighing of the criteria is usually felt to be the prerogative of those who established the technical committee. This implies that the relative importance of the criteria may have been assessed in the absence of immediate context, that is, in isolation from the alternatives. In general, we indeed assume that the relative importance of the criteria does not depend on the alternatives which happen to be present in the actual decision problem, but on certain emotional or social values in the decision maker's mind and on conditions which are (sometimes far) beyond the scope of the decision problem under consideration.

Although we recommend to use a small number of criteria only, we presented a number of examples in Chapter 4 to show that a large number of alternatives can sometimes easily be handled. This is particularly true if the performance of the alternatives can be expressed in physical or monetary units on a one-dimensional scale. One can also ask each of the technical experts to judge a limited number of alternatives only. The group should reasonably cover the complete set of alternatives, however. The evaluation of the criteria, which can technically be carried out in a similar manner, should be carried out completely by all decision makers. This is at least our experience in several projects with group decision making.

REFERENCES TO CHAPTER 5

1. Berlin, B., and Kay, P., *"Basic Color terms: Their Universality and Evolution"*. University of California Press, Berkeley, 1969.

2. Berthoz, A., *"Le Sens du Mouvement"*. Editions Odile Jacob, Paris, 1997.

3. Beutelbacher, A., and Petri, B., *"Der Goldene Schnitt"*. 2. Auflage. BI Wissenschafts-verlag, Mannheim, 1995.

4. Bruce, V., and Green, P.R., *"Visual Perception, Physiology, and Ecology"*. Lawrence Erlbaum, Hove, UK, 1990.

5. Bünning, E., *"The Physiological Clock, Circadian Rhythms and Biological Chronometry"*. The Heidelberg Science Library, Springer, New York, 1973.

6. Doczi, G., *"The Power of Limits, Proportional Harmonies in Nature, Art, and Architecture"*. Shambala, Boulder, Colorado, 1981.

7. Gibson, J.J., *"The Ecological Approach to Visual Perception"*. Houghton-Mifflin, Boston, 1979.

8. Hagenmaier, O., *"Der Goldene Schnitt, ein Harmoniegesetz und seine Anwendungen"*. Moos & Partner, Gräfelfing, Germany, 1984.

9. Hekkert, P., *"Artful Judgements"*. Doctoral Dissertation, Delft University of Technology, Delft, The Netherlands, 1995.

10.Herz-Fischler, R., *"A Mathematical History of Division in Extreme and Mean Ratio"*. Wilfrid Laurier University Press, Waterloo, Ontario, Canada, 1987.

11.Huntley, H.E., *"The Divine Proportion, a Study in Mathematical Beauty"*. Dover, New York, 1970.

12.Kay, P., and McDaniel, C.K., "The Linguistic Significance of the Meaning of Basic Color Terms". *Language* 54, 610 – 646, 1978.

13.Keeney, R., and Raiffa, H., *"Decisions with Multiple Objectives: Prefernces and value-Trade-Offs"*. Wiley, New York, 1976.

14.Kelso, J.A.S., *"Human Motor Behavior, an Introduction"*. Lawrence Erlbaum, Hillsdale, New Jersey, 1982.

15.Lipovetsky, S., and Lootsma, F.A., "Generalized Golden Sections, Repeated Bisections, and Aesthetic Pleasure". To appear in the *European Journal of Operational Research,* 1999.

16.Lootsma, F.A., *"Fuzzy Logic for Planning and Decision Making"*. Kluwer Academic Publishers, Boston, 1997.

17.Miller, G.A., "The Magical Number Seven Plus or Minus Two: Some Limits on our Capacity for Processing Information". *Psychological Review* 63, 81 – 97, 1956.

18.Peper, C.E., *"Tapping Dynamics"*. Doctoral Dissertation, Faculty of Human Movement Sciences, Free University, Amsterdam, 1995.

19.Rosenbaum, D.A., *"Human Motor Control"*. Academic Press, San Diego, California, 1991.

20.Roy, B., *"Méthodology Multicritère d'Aide à la Décision"*. Economica, Collection Gestion, Paris, 1985.

21.Runion, G.R., *"The Golden Section and Related Curiosa"*. Scott, Foresman and Company, Glenview, Illinois, 1972.

22.Taylor, J.R., *"Linguistic Categorization, Prototypes in Linguistic Theory"*. Clarendon Press, Oxford, 1995.

23.Wadum, J., "Vermeer in Perspective". In *"Johannes Vermeer"*, Mauritshuis, The Hague, and National Gallery of Art, Washington, DC. Waanders, Zwolle, The Netherlands, 67 – 79, 1995.

24. Winfree, A.T., *"The Timing of Biological Clocks"*. Scientific American Books, New York, 1987.

25. Winterfeldt, D. von, and Edwards, W., "*Decision Analysis and Behavioral Research*". Cambridge University Press, Cambridge, UK, 1986.

CHAPTER 6

GROUP DECISION MAKING

Although many decisions are made, not by individuals, but in groups such as boards, councils, and committees, the MCDA literature pays little attention to group decision making. Particularly the power relations in groups are overlooked. There is always a power game, however. In national decision-making bodies each member seems to have a weight which is proportional to the size of the organization which he/she represents. In international decision-making bodies the weights of the members are related to the population size or the Gross National Product of the respective countries. Alternatives which are weakly supported by the powerful members have therefore little chance of being adopted by the group, even when MCDA reveals a high degree of support for them under the erroneous assumption that all members have equal weights.

6.1 THE ANATOMY OF POWER

First, what is power? Galbraith (1983) defined it as the force that causes the persons subject to it to abandon their own preferences and to accept those of others. In general, there are three kinds of power.

1. Punitive power wins submission because it threatens the others with public rebuke, punishment, and/or violence, things which are physically or emotionally painful enough so that the others forgo their own preferences.

2. Compensatory power wins submission by incentives and/or rewards in the form of public praise, higher salaries, bonuses, etc. This type of power is thought to be far more civilized, far more consistent with the liberty and dignity of the individual, than punitive power.

3. Conditioned power wins submission by belief, education and/or indoctrination. Those exercising it and those subject to it are not always aware that it is being exerted. On the contrary, the acceptance of higher authority and the submission to the will of others becomes the higher preference of those submitting.

The sources of power are personality, property, and organization. Personality or leadership stands for a sweeping quality of intellect, moral certainty, physique, mind, or speech. Property or wealth accords an aspect of authority and a certainty of purpose. A high-ranking position in an organization gives access to all types of power. The sources of power are used in various mixtures. The power of the Church, for instance, was anciently supported by its access to punitive power in the present world and in the world to come, and by the compensatory attraction of its benefices. Overwhelmingly, however, its power depends on belief.

Power is pursued, not only for the services it renders to the interests of individuals and groups, but also for its own sake, for the emotional rewards inherent in the possession and exercise. The primary source of power today seems to be the disciplined organization such as a state bureaucracy or a modern corporation, where administrators and managers exercise their compensatory and conditioned power in the hierarchy. Decision-making teams in such an organization are confronted with the so-called bimodal symmetry: their external power depends on internal discipline. Hence, the need for consensus or a compromise which is accepted by all team members. If the team comes up with a majority report plus one or two minority statements, the deliberations will be resumed at other levels of the organization.

One can make a further distinction between formal power laid down in the assignment of tasks, responsibilities, and rules for weighted voting on the one hand, and informal power derived from power sharing, participation, and delegation on the other. Mintzberg (1983), who defined power as the capacity to affect the outcomes of an organization, therefore considered four power systems.

1. The System of Authority, the system of formal, legitimate power working via direct orders and decisions (personal control) and via standardized procedures for job execution and performance measurement (bureaucratic control).

2. The System of Ideology working via conditioning, belief, and indoctrination.

3. The System of Expertise working via the ability of experts, mainly analysts in staff departments, to cope with uncertainty.

4. The System of Politics, the system of illegitimate, unsanctioned, unauthorized power which is necessary to correct the deficiencies of the other systems, to ensure that all sides of an issue are fully debated, to bring the strongest members of the organization into leadership positions, and to promote organizational change.

Many power games originate from the attempts of line managers and analysts in staff departments to form particular coalitions. The analysts are not powerless. They are professionals, and they have the expertise to restructure the system of formal, legitimate power. They can replace personal control by bureaucratic control, see the development of information systems for production planning and inventory control.

Heller and Wilpert (1981) studied the process of power sharing at higher levels of modern organizations in the USA and in Western Europe. They analyzed the consequences of five styles of decision making in terms of improved quality of the decision, improved speed of decision making, as well as improved satisfaction with the decision process. The degree of power sharing, they found, is mainly determined by the percieved skills of the subordinates: more skills, more power.

In the present volume we limit ourselves to decision making in a public or a private bureaucracy, when the members have a common interest: they are supposed to present a joint compromise solution to those who asked the group to solve a given problem of choice. Not only the relative, formal power of the group members but also the criteria and their relative importance may have been prescribed to the group, albeit in vague terms which leave ample space for adjustment and interpretation. Furthermore, the power game is constrained by what is socially acceptable in the bureaucracy. We do not consider open conflicts in the initial phase when the actors use brute force and when an organizational structure for negotiations and problem solving is still absent. Similarly, we do not try to model the informal power which a group member may derive from his/her personality and skills.

To our knowledge this extension of MCDA is new. Spillman et al. (1980) as well as Aczél and Saaty (1983) propose methods to synthesize the pairwise comparisons expressed by the individual members of a group, but they do not weigh these statements. Thus they ignore a possible distinction between the group members on the basis of formal or informal power. Similarly, Hwang and Lin (1987) and Radford (1989), who typically work in the intersection of decision analysis and group decision making, avoid the issue of inhomogenuous power distributions in groups. One of the reasons may be that it is usually very difficult to obtain experimental evidence about decision-making processes in groups engaged in real-life situations in the day-to-day world.

Considering the Additive and the Multiplicative AHP, we assign a weight coefficient to each member of the group and we use these power coefficients in the logarithmic least-squares method whereby we analyze the pairwise-comparison matrices (see also Barzilai and Lootsma, 1997). We shall particularly be concerned with the special case that each member of the group judges every pair of alternatives under each of the criteria. This implies that the pairwise comparisons are complete: all cells in the pairwise-comparison matrices have exactly the same number of entries. Moreover, the normal equations associated with the weighted logarithmic least squares have an explicit solution: we can immediately find impact grades and scores for the alternatives by the calculation of arithmetic row means (Additive AHP) or geometric row means (Multiplicative AHP). Lastly, the order of the calculations is immaterial. It does not matter whether we first average over the entries within a cell to obtain a group opinion about a pair of alternatives and thereafter over a row within the matrix to obtain an impact grade or score for the corresponding alternative, or vice versa. At a higher level, when we aggregate the impact grades or scores of the alternatives in order to obtain final grades or terminal scores, we can also compute arithmetic means (Additive AHP) or geometric means (Multiplicative AHP) in an arbitrary order.

Mathematically, there is no difference between the behaviour of the power coefficients and the criterion weights, at least in the proposed aggregation procedure. They are used in an identical manner, both in the weighted arithmetic means and in the weighted geometric means. This seems to make sense. Aggregation over the criteria via the calculation of weighted arithmetic or geometric means is usually carried out in order to find a compromise between the conflicting preferential feelings within the mind of a single decision maker. Similarly, aggregation over the decision makers may be carried out in order to find a compromise between the conflicting preferential feelings within the group. The mathematical behaviour of the power coefficients therefore suggests that relative power is properly incorporated in the Additive and the Multiplicative AHP via the weighted logarithmic least-squares method. This is also illustrated by the power relations between the member countries of the European Union, at least under the assumption that the Gross National Product, the size of the population, or the number of seats in the European Parliament is a proper measure of relative power.

6.2 WEIGHTED JUDGEMENT

In the basic experiment at the first evaluation level of the analysis, two alternatives A_j and A_k are presented to decision maker d who is asked to express his/her graded comparative judgement under a particular criterion. We assume that the alternatives have unknown subjective values V_j and V_k which are the same for all decision makers in the group. Otherwise, consensus and/or a compromise would hardly be possible, whereas the group members are supposed to arrive jointly at a common group standpoint. This is a serious requirement (see also the bimodal symmetry mentioned in the previous section) which implies that the decision makers are requested to give up their individual preferences and to act primarily as group members. Particularly the French school in MCDA questions the existence of a coherent pattern of subjective values at the beginning of a decision process

(Roy and Vanderpooten, 1996). Even in the mind of a single decision maker such a pattern is not necessarily a psychological reality, but it might be created by the elicitation procedure itself. Nevertheless, many decisions are made within an organizational framework where the members have common values. Otherwise these members would not belong to the organization. This is discussed at length by Mintzberg (1983).

In the Multiplicative AHP the verbal comparative judgement, given by decision maker d in the above-named basic experiment and converted into a numerical value r_{jkd} according to the rules of Section 3.1, is taken to be an estimate of the ratio V_j/V_k. Furthermore,

$$\delta_{jkd} = {}^2\log r_{jkd}.$$

The symbol δ_{jkd} is the difference of grades which is used in the Additive AHP to represent the gradations of verbal comparative judgement. Next, we approximate the vector V of subjective values again via logarithmic regression. We introduce the set D_{jk} to denote the set of decision makers who actually expressed their opinion about the two alternatives under consideration, and we approximate the vector V of subjective values via the unconstrained minimization of the sum of squares

$$\sum_{j>k} \sum_{d \in D_{jk}} \left({}^2\log r_{jkd} - {}^2\log \upsilon_j + {}^2\log \upsilon_k \right)^2 p_d, \tag{6.1}$$

where the power coefficient p_d stands for the relative power of decision maker d. We take the power coefficients to be normalized so that they sum to unity. With the new variables

$$w_j = {}^2\log \upsilon_j$$

we can rewrite (6.1) as

$$\sum_{j>k} \sum_{d \in D_{jk}} \left(\delta_{jkd} - w_j + w_k \right)^2 p_d. \tag{6.2}$$

Note again that we minimize the sum of squares (6.2) regardless of whether we have ratio information (Multiplicative AHP) or difference information (Additive AHP). We obtain an optimal solution of (6.2) by solving the associated set of normal equations which is obtained by setting the first-order derivatives of (6.2) to zero. Using the properties

$$\delta_{jkd} = -\ \delta_{kjd} \text{ for any } j \text{ and } k,$$

and differentiating (6.2) with respect to w_j we find

$$\sum_{k=1}^{j-1} \sum_{d \in D_{jk}} \left(\delta_{jkd} - w_j + w_k \right) p_d - \sum_{k=j+1}^{n} \sum_{d \in D_{kj}} \left(\delta_{kjd} - w_k + w_j \right) p_d = \sum_{\substack{k=1 \\ k \neq j}}^{n} \sum_{d \in D_{jk}} \left(\delta_{jkd} - w_j + w_k \right) p_d = 0,$$

so that the normal equations themselves take the form

$$R_j w_j - \sum_{\substack{k=1 \\ k \neq j}}^{n} C_{jk} w_k = \sum_{\substack{k=1 \\ k \neq j}}^{n} \sum_{d \in D_{jk}} \delta_{jkd} p_d, \; j = 1, \ldots, n, \tag{6.3}$$

where the symbol

$$R_j = \sum_{\substack{k=1 \\ k \neq j}}^{n} \sum_{d \in D_{jk}} p_d$$

denotes the total amount of power in row j of the matrix of pairwise comparisons, whereas the symbol

$$C_{jk} = \sum_{d \in D_{jk}} p_d$$

stands for the total amount of power in cell (j, k) of the matrix. The normal equations are dependent. There is at least one additive degree of freedom in the unconstrained minima of the function (6.2) because there are only differences of variables in the sum of squares. Hence, there is at least one multiplicative degree of freedom in the unconstrained minima of the function (6.1). We can therefore only draw conclusions from differences $w_j - w_k$ and from ratios υ_j/υ_k.

When each decision maker expresses his/her opinion about every pair of alternatives, then

$$D_{jk} = \{1, \ldots, G\} \text{ for any } j \text{ and } k,$$

where G represents the size of the group. Since the power coefficients are normalized the normal equations (6.3) can be reduced to

$$n w_j - \sum_{k=1}^{n} w_k = \sum_{k=1}^{n} \sum_{d=1}^{G} \delta_{jkd} p_d, \quad j = 1, \ldots, n,$$

if we take $\delta_{jjd} = 0$ for all j. We use again the additive degree of freedom in the solutions to the normal equations in order to set the sum of the variables to zero, whence

$$w_j = \frac{1}{n} \sum_{k=1}^{n} \sum_{d=1}^{G} \delta_{jkd} p_d. \tag{6.4}$$

This expression yields the impact grades for the Additive AHP under the criterion which we consider (possibly after a shift so that the grades are conveniently situated between 4 and 10). An unnormalized solution to the weighted logarithmic least squares problem of minimizing (6.1) can now explicitly be written as

$$v_j = \sqrt[\eta]{\prod_{k=1}^{n} \prod_{d=1}^{G} r_{jkd}^{p_d}}. \tag{6.5}$$

These are the unnormalized impact scores for the Multiplicative AHP. Obviously, we have to calculate two arithmetic means for the Additive AHP and two geometric means for the Multiplicative AHP. These operations can be carried out in any order. When we average over row j in the pairwise-comparison matrix of decision maker d, we obtain his/her estimates of the subjective values of the alternatives. When we average over the decision makers in cell (j, k), we obtain the group's preference for A_j with respect to A_k. These two operations are combined in the formulas (6.4) and (6.5).

We are now in a position to consider the role of the criteria. We start from the assumption that the (normalized) criterion weights c_i of the respective criteria C_i, $i = 1,\ldots, m$, have been set in previous group discussions and that they are unanimously accepted by the members. Alternatively, these weights were already given at the time when the decision-making group was established. Let us now consider the basic experiment where decision maker d judges the alternatives A_j and A_k under criterion C_i. We take the symbol r_{ijkd} to represent the numerical value assigned to his/her verbal estimate of V_{ij}/V_{ik}, the ratio of the subjective values of the alternatives under consideration. The impact scores s_{ij} of the alternatives A_j, $j = 1,\ldots, n$, under the respective criteria C_i, $i = 1,\ldots, m$, follow from geometric-mean calculations over the decision makers and over the rows of the pairwise-comparison matrices. Lastly, we use the geometric-mean aggregation rule of the Multiplicative AHP to obtain an unnormalized terminal score t_j for alternative A_j according to the expression

$$t_j = \prod_{i=1}^{m} s_{ij}^{c_i}$$

or, equivalently,

$$t_j = \sqrt[\eta]{\prod_{i=1}^{m} \prod_{k=1}^{n} \prod_{d=1}^{G} r_{ijkd}^{c_i p_d}}. \tag{6.6}$$

Moreover, using the features of logarithmic coding, we have the symbol

$$\delta_{ijkd} = {}^{2}\log r_{ijkd}$$

to represent the difference of grades employed in the Additive AHP. The final grade of the alternative A_j according to the Additive AHP is therefore given by the expression

$$f_j = \frac{1}{n}\sum_{i=1}^{m}\sum_{k=1}^{n}\sum_{d=1}^{G} c_i p_d \delta_{ijkd}. \tag{6.7}$$

Obviously, we have a sequence of geometric means (formula (6.6)) or arithmetic means (formula (6.7)) which can be computed in any order. In addition, the intermediate results make sense whatever the order of the computations is: averaging over the rows in the pairwise-comparison matrices of the respective decision makers and thereafter over the criteria, for instance, produces the final grades (Additive AHP) or the terminal scores (Multiplicative AHP) of the alternatives for each decision maker individually. The original AHP does not have this flexibility, see also Section 3.6.

The above ideas can immediately be applied in SMART. Let g_{ijd} stand for the grade assigned under criterion C_i to alternative A_j by decision maker d. The final grade of alternative A_j can accordingly be written as

$$f_j = \sum_{i=1}^{m}\sum_{d=1}^{G} c_i p_d g_{ijd}. \tag{6.8}$$

Here we only have to apply two arithmetic-mean calculations in an arbitrary order. Note that we have one set of criterion weights and one set of power coefficients only, regardless of whether we turn to the Additive AHP, the Multiplicative AHP, or SMART.

6.3 SCALE VALUES FOR RELATIVE POWER

Because power coefficients behave mathematically like criterion weights, we can assign numerical scale values to verbal qualifications such as *somewhat more, definitely more, much more,* and *vastly more* powerful in the same way as we quantified verbal statements such as *somewhat more,...., vastly more* important, see Section 2.5 and Section 4.2. Two decision makers which are equal in power will have power coefficients with the ratio 1:1. We now try to establish the other end of the range, the ratio which corresponds to *vastly more* power, in the sense that one of the two decision makers completely overrules the other. Consider a decision-making group with two members D_1 and D_2, and suppose that there are two real or imaginary alternatives A_j and A_k such that the decision makers have an equally strong but inverse preference for them under a particular criterion. These preferences are estimated by

$$2^{\delta_{jk}} \text{ and } 2^{-\delta_{jk}}$$

respectively, where δ_{jk} designates the selected gradation of D_1's comparative judgement. We assume that these preferences do not depend on the performance of the alternatives under the remaining criteria. Now, imagine that the group of the two decision makers jointly has a preference for A_j versus A_k estimated by

$$2^{\vartheta_{jk}},$$

where ϑ_{jk} denotes the gradation which designates the group's preference. Taking π to stand for the relative power (the ratio of the power coefficients) of D_1 with respect to D_2, we obtain by the geometric-mean aggregation rule, and henceforth by the addition of the logarithms, that

$$\frac{\pi}{\pi+1}\delta_{jk} - \frac{1}{\pi+1}\delta_{jk} = \vartheta_{jk},$$

whence

$$\pi = \frac{\delta_{jk} + \vartheta_{jk}}{\delta_{jk} - \vartheta_{jk}}.$$

It will be clear that $|\vartheta_{jk}| < |\delta_{jk}|$ because the group preference cannot absolutely exceed the preference expressed by the decision makers individually. It is easy to verify now that π varies roughly between $^1/16$ and 16 when δ_{jk} varies over the integer values between -8 and 8 and ϑ_{jk} over the integer values between $-|\delta_{jk}|$ and $|\delta_{jk}|$. The rather extreme value $\pi = 15$ is obtained when $\delta_{jk} = 8$ and $\vartheta_{jk} = 7$, which means that D_1's *very strong* preference for A_j with respect to A_k almost completely wipes out the equally strong but inverse preference of D_2. Thus we take the ratio 16:1 to roughly stand for a *vastly more* powerful position, at least when we consider two decision makers only. This is also a plausible ratio, however, when we consider three or more decision makers. When D_1 has a *very strong* preference for A_j with respect to A_k ($\delta_{jk} = 8$), for instance, whereas two other decision makers D_2 and D_3 have a *definite* but inverse preference ($\delta_{jk} = -4$), then a joint preference designated by $\vartheta_{jk} = 7$ is reached if the power coefficient of D_1 has the ratio 22:1 with respect to the coefficients of D_2 and D_3. Other extreme cases confirming that the ratio 16:1 is a rough but reasonable representation of *vastly more* power can easily be constructed.

A simple geometric sequence of values between 1 and 16 with 5 echelons, corresponding to *equal, somewhat more, definitely more, much more*, and *vastly more* power, is given by the sequence with the progression factor 2. Thus, we obtain the geometric scale 1, 2, 4, 8, 16 to designate the gradations of relative power. If we also allow for hesitations between two adjacent gradations of relative power, we eventually have a geometric sequence of scale values with the progression factor $\sqrt{2}$. The inverse echelons 1, $^1/2$, ... , $^1/16$ are of course taken to designate *equal,..., vastly less* power.

In summary, we found that there is a distinction between the scaling of the relative power of the decision makers and the relative importance of the criteria on the one hand and the scaling of the relative preference for the alternatives on the other. The progression factors are different: $\sqrt{2}$ for relative power and relative importance, and 2 for relative preference.

It is worth noting here that the elicitation of criterion weights and the assignment of power coefficients are politically more delicate than the performance evaluation of the alternatives. The choice of the criteria and the vague verbal formulation of their relative importance are usually felt to be the prerogative of those who established the decision-making group. In weighted-judgement procedures power is unequally distributed via the formal assignment of weights to the members of the group. Generally, however, relative power is an issue that cannot even be discussed in the group although it is observable throughout the decision process.

A simple example to illustrate the possible significance of the ratio 16:1 assigned to *vastly more* power is given by the relations between the member countries of the European Union. Table 6.1 shows their population size and their Gross Domestic Product.

Table 6.1 *Population, Gross Domestic Product, and number of seats in the European Parliament of the* 12 *member states of the European Union in* 1993. *Data from the Eurostat Yearbook* 1995.

	Population 1993 in millions	GDP 1993 in 10^9 ECU	Number of Seats in European Parliament
Belgium	10.07	125	25
Denmark	5.18	85	16
Germany	80.98	1105	99
Greece	10.35	59	25
Spain	39.05	277	64
France	57.53	810	87
Ireland	3.56	37	15
Italy	56.96	658	87
Luxembourg	0.40	8	6
The Netherlands	15.24	205	31
Portugal	9.87	41	25
United Kingdom	58.10	709	87
European Union	347.29	4119	567

According to both yardsticks, Germany is leading. In a dispute with Germany, when the GDP is the proper indicator for the respective power positions in the Union, the countries with a GDP below $1105/16 \approx 70$ billion ECU are likely to be ignored in applications of MCDA with weighted judgement, unless they are jointly members of a stronger coalition. Similarly, in conflicts with Germany when the population size is the overt indicator, the countries with less than $81/16 \approx 5$ million inhabitants are likely to have no influence in applications of MCDA with weighted judgement. Of course, conflicts are usually much more complicated and a single indicator is not always enough to explain the relative power positions of the parties concerned. The number of seats per country in the European Parliament, however, may be a one-dimensional representation of the balance of power which eventually emerged after many years of wheeling and dealing. Is it pure coincidence that the ratio of the largest delegation (Germany, 99) to the smallest one (Luxembourg, 6) is roughly 16:1, so that the smallest country is totally overruled in extreme cases only?

Let nobody have the illusion that the above incorporation of power relations would be welcome in practical applications of MCDA with weighted judgement. In a project carried out to the order of Directorate-General XII (Science, Research, and Development) of the European Union (Lootsma, Mensch, and Vos, 1990, see also Chapter 7), our proposal to weigh the judgemental statements of the participating decision makers by the population size or the GDP of the respective countries was immediately rejected. One of the reasons may have been that unanimity would be required in the decisions at hand. Moreover, it was presumably too early to model the on-going power game in the Union explicitly via MCDA, although it was already possible with an earlier version of our REMBRANDT programme (L. Rog, Delft University of Technology). In other projects we have similar experiences. Although public administrators and industrial managers privately agree that it could be interesting to explore the relative power of various coalitions in a decision-making body, an open analysis of the power game seems to be too hazardous in actual decision processes.

The prospects for weighted judgement are far from hopeless, however, because weighted voting is not uncommon. In the Council of Ministers of the European Union, for instance, although unanimity is required for all major and most routine decisions, a qualified majority is sufficient in certain prespecified cases. In 1993, the larger states were entitled to more votes than the smaller ones: France, Germany, Italy, and the United Kingdom 10; Spain, 8; Belgium, Greece, the Netherlands, and Portugal 5; Denmark and Ireland 3; Luxembourg 2. Bueno de Mesquita and Stokman (1994) analyzed the decision processes of the Union in four issue areas, using the voting power to model the relative power of the member states. An interesting additional feature of their approach was the salience factor which models the willingness of the member states to withhold their power and to save it for other issues about which they care more, or for log-rolling, an arrangement between group members by which some of them agree to support the position of others in one decision in return for mutual support in another situation. This aspect — how to handle power in a fair and/or subtle manner and when to abstain from using it — has not been considered here.

It will be obvious that weighted judgement, as it is described here, is not the same as weighted voting. A simple example may be sufficient to illustrate the issue. In a decision-making body consisting of three members with the respective weights 3, 1, and 1, weighted voting would be totally unattractive for the second and the third member because of the dictatorial position of the first member. Weighted judgement, however, would still be feasible because the decisions can be influenced by the second and the third member, despite the strong position of the first member. For a simple and detailed description of weighted voting we refer the reader to Taylor (1995), who also presents a mathematical model for power which is based upon the number of situations where a member can just tip the balance between two coalitions with opposite standpoints.

6.4 THE HOUSE SELECTION PROBLEM

The problem of how to select a house is encountered by almost anybody, whether he/she is a decision maker or an analyst. Hence, it is an interesting test case for analysts. It enables them to see a decision problem from the viewpoint of a decision maker. Moreover, it is usually a group decision problem, although the group has the smallest possible size. It consists of two members only: husband and wife. The children may affect the choice, but in the present chapter we leave them out of consideration. Their interests are taken into account via the choice of the criteria and the assignment of criterion weights. In addition, this problem enables us to illustrate how the search for a compromise solution is affected by the question of when to compromise: throughout the analysis, or at the end of the analysis, when the decision makers have individually made up their mind. This procedural question emerges when there is no agreement on the weights of the criteria, something that we ignored in the previous sections. Acceleration of the convergence towards a compromise solution may also be obtained via the power game. Initially, we shall be assuming that the two partners are equally powerful, but we shall gradually deviate from this assumption. In real-life situations it may also easily happen that one of the two partners tacitly assigns more power or influence to the other.

We shall be concerned here with the choice of a house in a medium-size town surrounded by woods and arable land. Dominated by the university, the lively down-town area is full of shops, restaurants, cinemas, theatres, and offices. The husband, a senior scientist, works in an agricultural laboratory, some twelve miles from the town centre. His wife will easily find a job in a down-town office as soon as the children are grown up. After a thorough screening, the spouses decide to consider the following four options:

1. Liberty Street, $125,000, a detached house built in the first decade of this century in the down-town area. It is in a poor condition, but it is spacious and the original style is still attractive. It is roughly one mile away from the Pioneer High School selected for the children of the family, but twelve miles from the laboratory.

2. Packard Road, $160,000, a semi-detached house in a new condominium, roughly four miles from the town centre and from the Pioneer High School, but fifteen miles from the laboratory.

3. Huron River Drive, \$200,000, a villa built in the nineteen fifties. It is in a good condition, located in a scenic area, roughly five miles from the town centre, the Pioneer High School, and the laboratory.

4. Dexter Lane, \$250,000, a nicely renovated nineteenth-century farm house located in a rural area, roughly eight miles from the town centre and the Pioneer High School, but only two miles from the laboratory.

In the screening phase the spouses discover that they intuitively employ the same criteria: actual price, quality of the location, style and condition of the house, distance from the school, and distance from the laboratory. They have quite different views, however, on the relative importance. An analysis via pairwise comparisons brings up the matrices of Table 6.2 and Table 6.3. Using the arithmetic row means as exponents of the progression factor $\sqrt{2}$ and normalizing the results thereafter one can easily obtain the criterion weights according to the method described in Section 3.3. Note that the spouses decided to judge the houses on the basis of the actual price and the actual condition. With an open eye for possibilities or affordances (Section 5.2) they could also take the desired condition and the actual price plus the estimated renovation cost as performance criteria. What matters in such a decision is the end result: the total price, the style of the house, and the life-style which it eventually affords. For the time being, however, the actual situation is taken to be decisive.

Table 6.2 *Husband's pairwise-comparison matrix with estimated differences of grades assigned to pairs of criteria, and husband's normalized criterion weights*

	Price	Location	Condition	School	Laboratory	Weight
Price	0	-2	+2	+4	+4	0.256
Location	+2	0	+3	+6	+6	0.478
Condition	-2	-3	0	+2	+2	0.137
School	-4	-6	-2	0	0	0.064
Laboratory	-4	-6	-2	0	0	0.064

Table 6.3 *Wife's pairwise-comparison matrix with estimated differences of grades assigned to pairs of criteria, and wife's normalized criterion weights.*

	Price	Location	Condition	School	Laboratory	Weight
Price	0	+3	+1	+2	+4	0.356
Location	-3	0	-2	-1	+1	0.126
Condition	-1	+2	0	+1	+3	0.252
School	-2	+1	-1	0	+2	0.178
Laboratory	-4	-1	-3	-2	0	0.089

It is possible for the spouses to find compromise criterion weights now, but they may also postpone the search for a compromise solution until the evaluation has been completed. Table 6.4 shows compromise criterion weights for various ratios of power coefficients (relative power or influence), to be used in the next section on the assumption that the husband leaves the final decision increasingly to his wife.

Table 6.4 *Compromise criterion weights for various ratios of power coefficients assigned to husband and wife.*

	H:W = 1:1	H:W = 1:2	H:W = 1:4
Price	0.330	0.344	0.351
Location	0.268	0.212	0.173
Condition	0.203	0.222	0.235
School	0.117	0.137	0.153
Laboratory	0.082	0.086	0.080

Let us now look at the evaluation of the alternatives under the respective criteria. The spouses largely agree in their assessment of the houses under the criteria of price, distance from the school, and distance from the laboratory. With the range of acceptable performance between $120,000 and $500,000 for the price, between 0 and 16 miles for the distance from the school, and between 0 and 32 miles for the distance from the laboratory, they judged the houses in pairs as shown in the Tables 6.5 – 6.7. Taking the arithmetic row means as exponents of the progression factor 2 and normalizing the results thereafter, one obtains the impact scores of the Multiplicative AHP which are displayed in the last column of the respective Tables.

Table 6.5 *Pairwise-comparison matrix with estimated differences of grades assigned to pairs of houses under the price criterion, and normalized AHP impact scores.*

	Liberty	Packard	Huron	Dexter	Scores
Liberty	0	+2	+3	+4	0.696
Packard	-2	0	+1	+2	0.174
Huron	-3	-1	0	+1	0.087
Dexter	-4	-2	-1	0	0.043

Table 6.6 *Pairwise-comparison matrix with estimated differences of grades assigned to pairs of houses under the criterion of distance from the school, and normalized AHP impact scores.*

	Liberty	Packard	Huron	Dexter	Scores
Liberty	0	+2	+2	+3	0.615
Packard	-2	0	0	+1	0.154
Huron	-2	0	0	+1	0.154
Dexter	-3	-1	-1	0	0.077

Table 6.7 *Pairwise-comparison matrix with estimated differences of grades assigned to pairs of houses under the criterion of distance from the laboratory, and normalized AHP impact scores.*

	Liberty	Packard	Huron	Dexter	Scores
Liberty	0	+1	-2	-3	0.080
Packard	-1	0	-2	-3	0.056
Huron	+2	+2	0	-2	0.226
Dexter	+3	+3	+2	0	0.638

The spouses strongly disagree in their assessment when they consider the quality of the location and the condition of the houses (which also includes an assessment of the style). Instead of pairwise comparisons they apply direct rating. However, using the grades as exponents of the progression factor 2 and normalizing the results, they can immediately obtain the AHP impact scores so that an integrated assessment of the houses is still feasible. The Tables 6.8 and 6.9 show the grades and the impact scores of husband and wife individually under the respective criteria, as well as compromise impact scores with varying ratios of power coefficients.

Table 6.8 *Direct assessment of the houses under quality of the location, impact scores of husband and wife individually, and compromise impact scores for varying ratios of power coefficients.*

	Husband		Wife		Compromise impact scores		
	Grades	Scores	Grades	Scores	H:W = 1:1	H:W = 1:2	H:W = 1:4
Liberty	4	0.024	7	0.182	0.076	0.105	0.133
Packard	6	0.098	8	0.364	0.214	0.264	0.305
Huron	9	0.780	8	0.364	0.604	0.527	0.462
Dexter	6	0.098	6	0.091	0.107	0.105	0.101

Table 6.9 *Direct assessment of the houses under condition and style, impact scores of husband and wife individually, and compromise impact scores for varying ratios of power coefficients.*

	Husband		Wife		Compromise impact scores		
	Grades	Scores	Grades	Scores	H:W = 1:1	H:W = 1:2	H:W = 1:4
Liberty	4	0.019	6	0.083	0.056	0.069	0.077
Packard	6	0.075	9	0.667	0.315	0.437	0.536
Huron	8	0.302	7	0.167	0.315	0.275	0.233
Dexter	9	0.604	6	0.083	0.315	0.219	0.154

Table 6.10 and Table 6.11 show the terminal scores assigned by husband and wife individually to the houses in question. Obviously, there is a considerable amount of disagreement. The husband clearly prefers the villa on Huron River Drive, mainly because of the location. His wife prefers the detached house in Liberty Street, mainly because of the price, but she might easily be persuaded to accept the semi-detached house in Packard Road because of its condition and style. The conflict is not hopeless, however. In the next section, we will explore the potential of various techniques to find a compromise solution.

Table 6.10 *Husband's terminal scores for the houses under consideration.*

Criteria	Weights	Liberty	Packard	Huron	Dexter
Price	0.256	0.696	0.174	0.087	0.043
Location	0.478	0.024	0.098	0.780	0.098
Condition	0.137	0.019	0.075	0.302	0.604
School	0.064	0.615	0.154	0.154	0.077
Laboratory	0.064	0.080	0.056	0.226	0.638
Terminal scores		0.119	0.175	0.524	0.182

Table 6.11 *Wife's terminal scores for the houses under consideration.*

Criteria	Weights	Liberty	Packard	Huron	Dexter
Price	0.356	0.696	0.174	0.087	0.043
Location	0.126	0.182	0.364	0.364	0.091
Condition	0.252	0.083	0.667	0.167	0.083
School	0.178	0.615	0.154	0.154	0.077
Laboratory	0.089	0.080	0.056	0.226	0.638
Terminal scores		0.375	0.319	0.199	0.106

6.5 SEARCHING A COMPROMISE SOLUTION

The search for a compromise solution may start in an early phase of the analysis, as soon as there are contradictory results. In the actual house selection problem, for instance, the search may start with the generation of compromise criterion weights under the assumption that the two partners are equal in power or influence. These compromise weights are shown in the second column of Table 6.4. The search for a compromise solution continues during the evaluation of the houses under the criteria of location and condition. The compromise impact scores under the assumption of equal power may be found in Table 6.8 and Table 6.9. With these weights and scores one eventually obtains the compromise terminal scores of Table 6.12. Note that the results are due to the search for a compromise solution throughout the analysis.

Table 6.12 *Searching a compromise solution throughout the analysis, with the ratio* 1:1 *of the power coefficients. Compromise terminal scores for the four houses under consideration.*

Criteria	Weights	Liberty	Packard	Huron	Dexter
Price	0.330	0.696	0.174	0.087	0.043
Location	0.268	0.076	0.214	0.604	0.107
Condition	0.203	0.056	0.315	0.315	0.315
School	0.117	0.615	0.154	0.154	0.077
Laboratory	0.082	0.080	0.056	0.226	0.638
Compromise throughout analysis		0.269	0.264	0.311	0.156

Obviously, the villa on Huron River Drive seems to be the preferred solution. It is also the preferred solution of one of the spouses. In general, however, compromises throughout the analysis may bring up a solution which ranks second or lower for both partners individually.

The search for a compromise solution may also be postponed until the end of the analysis. One only has to wait until the terminal scores of the two partners individually are known, and thereafter one calculates a compromise solution which is a geometric mean, possibly normalized, of the terminal scores. The results are shown in Table 6.13.

Table 6.13 *Searching a compromise solution at the end of the analysis, with the ratio* 1:1 *of the power coefficients. Compromise terminal scores for the four houses under consideration.*

	Liberty	Packard	Huron	Dexter
Husband's terminal scores	0.119	0.175	0.524	0.182
Wife's terminal scores	0.375	0.319	0.199	0.106
Compromise at end of analysis	0.232	0.260	0.355	0.153

Again, the villa on Huron River Drive emerges as the preferred solution. Note that the compromise terminal scores in the Tables 6.12 and 6.13 are not identical because there is no agreement on the criterion weights. However, as soon as a set of criterion weights has unanimously been accepted by the two partners, the terminal scores do not depend on the order of the computations (Section 6.2). In such a situation the terminal scores obtained via compromises throughout the analysis coincide with the terminal scores obtained at the end of the analysis.

A group may assign more power or influence to those members who are really unhappy with the proposed compromise solution which emerges via the analysis. Let us suppose here that the husband is increasingly willing to leave the final decision to his wife. Thus, he may accept that power coefficients with the ratio 1:2, or even 1:4, are assigned to their comparative judgement. The solutions resulting from the searches for a compromise throughout and at the end of the analysis are shown in Table 6.14. We omit a detailed display of the calculations here. It will be obvious how the terminal scores have been obtained via compromises throughout the analysis (use the columns with the appropriate ratio H:W in the respective Tables). The terminal scores at the end of the analysis are weighted geometric means (possibly normalized) of the terminal scores of husband and wife individually (Table 6.10 and Table 6.11).

Table 6.14 *Compromise terminal scores depending on the procedure (compromise throughout the analysis or at the end of the analysis) under varying ratios of power coefficients.*

Compromise	H:W	Liberty	Packard	Huron	Dexter
throughout the analysis	1:1	0.269	0.264	0.311	0.156
at the end of the analysis	1:1	0.232	0.260	0.355	0.153
throughout the analysis	1:2	0.312	0.285	0.264	0.140
at the end of the analysis	1:2	0.279	0.284	0.299	0.138
throughout the analysis	1:4	0.341	0.299	0.234	0.126
at the end of the analysis	1:4	0.317	0.300	0.257	0.125

Note the shift from preference for the villa on Huron River Drive to preference for the detached house in Liberty Street. Under the power ratio 1:2 the preferred solution depends on the procedural question of when to compromise: throughout the analysis or at the end of the analysis. Under the power ratio 1:4 the preferred solution is the house in Liberty Street, regardless of the procedure.

The observation that a compromise solution may be in conflict with the individual preferences of the members of a group is due to Ramanathan and Ganesh (1994). They presented an example to show that an alternative which is preferred by all members individually may rank second or lower in the compromise solution. Van den Honert and Lootsma (1996) demonstrated that this is due to the procedure. The search for a compromise solution throughout the analysis is based upon the assumption that the group acts as a single decision maker. The search at the end of the analysis is likely to occur when the group is merely a collection of individuals.

The above calculations have been carried out with the REMBRANDT program (L. Rog, Delft University of Technology) for MCDA in groups with inhomogenuous power distributions. REMBRANDT generates compromise weights and scores throughout the analysis because in the nineteen eighties we had many projects where the pairwise comparisons were incomplete. On several occasions there was not enough information to calculate the terminal scores for each decision maker individually (see also Section 4.3).

Thus, there was no choice: we had to take the group as a single decision maker because several group members did not really "cover" the criteria and/or the alternatives.

6.6 PROSPECTS FOR MCDA IN GROUPS

In the public and in the private sector many decisions are made in groups (boards, councils, management teams, etc.), even when the initiatives come from individual members with creative views and a strong orientation towards the future. Group decisions are necessary when administrators and managers have joint responsibilities. Group decisions may also be pursued to conceal individual responsibilities. The literature on MCDA, however, seems to concentrate on methods for individual decision makers with a large amount of autonomy. So, what are the prospects for MCDA beyond the support for individuals? If MCDA cannot be used in groups, will it merely remain a hobby instead of a widely used tool for management?

Let us not discuss the question in general, but let us consider the prospects for MCDA from the viewpoint of the chairperson in a decision-making body (Lootsma, 1998). Although he/she is frequently considered to be a *primus inter pares* only, there are certain obligations which cannot be delegated to other members. He/she usually controls the decision-making procedures in the group, and he/she is usually the unique member to communicate with the outside world on behalf of the group. On many occasions the chairperson has been appointed, not by the group itself, but by higher authorities who originally established the group. Thus, the chairperson is more or less responsible for the recommendations and the decisions of the group. He/she has to be aware of the hidden agenda of the higher authorities and of the reasons why he/she happens to be in the chair. It cannot therefore be surprising that the chairperson, aware of the conflicting views and opinions in the group, should tend to avoid stumbling blocks such as the precise description of performance criteria and the elicitation of their weights. It is sometimes easier for the group to agree on a preferred alternative, perhaps for different reasons, than to agree on the weights to assign to the various criteria (Walker et. al., 1994). The key question now is: under which conditions could the chairperson propose to use the formalized procedures of MCDA, possibly supported by electronic devices such as a decision room with networked PC's and a public screen for brainstorming and voting?

The proposal will not alleviate the chairperson's task. He/she is responsible for many administrative details: the group has to be present at a given time, the hardware and the software must be operational, cooperation with the facilitator in the group decision room must have been tested beforehand, and the system has to be foolproof because there are always group members who try to amplify their influence by an improper (?) utilization of the MCDA methods. Timing is also important. The group must be prepared to come to a conclusion. This is not always the case, however. There are members who delay a decision until there is only one alternative option left. Others defer a decision to higher authorities. Those who typically weigh the pros and cons of various alternatives may constitute a minority.

The chairperson takes a risk when he/she proposes to follow the formalized approach of MCDA. If the group falls apart into quarelling fractions, it is usually taken to be his/her responsibility. In general, one cannot expect a fragmented, incoherent group to cooperate successfully as soon as a formalized procedure has been imposed (a dentist does not put a golden crown on a rotten root either).

Would MCDA be a tool for a charismatic chairperson only? The question is irrelevant because there are not many charismatic leaders. In order to operate successfully with MCDA in a decision room, the chairperson must be determined to make a unit out of the group, and the group has to feel that this is the objective of the chairperson. In addition, the group has to be aware of the bimodal symmetry (Galbraith, 1983): external power depends on internal discipline. Closed ranks only, no majority report plus one or two minority reports. Hence, the group has to arrive at a consensus or a compromise which is accepted by all members. That is the agenda for a successful group decision, possibly supported by MCDA.

The possible benefit of MCDA is that the criteria become operational. Although the members may remain anonymous in the electronic brainstorming and voting procedures, the criterion weights are openly shown, and this may lead to deep-going discussions and eventually to an agreement at a deeper level than one might initially expect. The technology of a decision room will also eliminate the advantages of certain discussion techniques. The members with strong verbal skills who usually dominate a meeting lose their grip on the silent majority as soon as the buttons are to be pressed. That may explain why some people dislike the procedures in a decision room.

A double-edged sword is the power shift that may be observed in a decision room (LaPlante, 1993). The anonymous procedures promote an egalitarian attitude. This may be a stumbling block in more authoritarian cultures (France, Japan, see Hofstede, 1994), where decisions are usually deferred to the boss. On the other hand, if the chairperson happens to be the boss, then he/she may welcome the anonymous procedures and the technological sophistication of a decision room because they reveal the hidden feelings in the group.

REFERENCES TO CHAPTER 6

1. Aczél, J., and Saaty, Th.L., "Procedures for Synthesizing Ratio Judgements". *Journal of Mathematical Psychology* 27, 93 – 102, 1983.

2. Barzilai, J., and Lootsma, F.A., "Power Relations and Group Aggregation in the Multiplicative AHP and SMART". *Journal of Multi-Criteria Decision Analysis* 6, 155 – 165, 1997. In the same issue there are comments by O.I. Larichev (p. 166), P. Korhonen (pp. 167 – 168), and L.G. Vargas (pp. 169 – 170), as well as a response by F.A. Lootsma and J. Barzilai (pp. 171 – 174).

3. Bueno de Mesquita, B., and Stokman, F.N., *"European Community Decision Making"*. Yale University Press, New Haven, CT, 1994.

4. Galbraith, J.K., *"The Anatomy of Power"*. Houghton-Mifflin, Boston, 1983.

5. Heller, F.A., and Wilpert, B., *"Competence and Power in Managerial Decision Making"*. Wiley, New York, 1981.

6. Hofstede, G., "Management Scientists are Human". *Management Science* 40, 4 – 13, 1994.

7. Honert, R. van den, and Lootsma, F.A., "Group Preference Aggregation in the Multiplicative AHP. The Model of the Group Decision Process and Pareto Optimality". *European Journal of Operational Research* 96, 363 – 370, 1996.

8. Hwang, C.L., and Lin, M.J., *"Group Decision Making under Multiple Criteria"*. Springer, Berlin, 1987.

9. LaPlante, A., "Nineties Style Brainstorming". *Technology Supplement to Forbes Magazine*, October 25, pp. 44 – 61, 1993.

10. Lootsma, F.A., "Prospects for MCDA in Groups". *Journal of Multi-Criteria Decision Analysis* 7, 121 – 122, 1998.

11. Lootsma, F.A., Mensch, T.C.A., and Vos, F.A., "Multi-Criteria Analysis and Budget Reallocation in Long-Term Research Planning". *European Journal of Operational Research* 47, 293 – 305, 1990.

12. Mintzberg, H., *"Power in and around Organizations"*. Prentice-Hall, Englewood Cliffs, NJ, 1983.

13. Radford, K.J., *"Individual and Small Group Decisions"*. Springer, New York, 1989.

14. Ramanathan, R., and Ganesh, L.S., "Group Preference Aggregation Methods employed in the AHP: an Evaluation and an Intrinsic Process for Deriving Members' Weightages". *European Journal of Operational Research* 79, 249 – 265, 1994.

15. Roy, B., and Vanderpooten, D., "The European School of MCDA: Emergence, Basic Features, and Current Works". *Journal of Multi-Criteria Decision Analysis* 5, 22 – 37, 1996. In the same volume there is a comment by F.A. Lootsma (pp. 37 – 39) as well as a response by B. Roy and D. Vanderpooten (pp. 165 – 166).

16. Spillman, B., Spillman, R., and Bezdek, J., "A Fuzzy Analysis of Consensus in Small Groups". In P.P. Wang and S.K. Chang (eds.), *"Fuzzy Sets, Theory and Applications to Policy Analysis and Information Systems"*. Plenum, New York, 1980, pp. 291 – 308.

17. Taylor, A.D., *"Mathematics and Politics"*. Springer, New York, 1995.

18. Walker, W., Abrahamse, A., Bolten, J., Kahan, J.P., Riet, O. van de, Kok, M., and Braber, M. den, "A Policy Analysis of Dutch River Dike Improvements: Trading Off Safety, Cost, and Environmental Impacts". *Operations Research* 42, 823 - 836, 1994.

CHAPTER 7

RESOURCE ALLOCATION

Since the final grades (SMART, Additive AHP) and the terminal scores (Multiplicative AHP) provide global preference information in a cardinal form they may be used to allocate scarce resources to the alternatives under consideration. That is at least the working hypothesis of the present chapter. The feasibility of the approach depends largely on the sensitivity of the results. The proposed allocation should be robust, that is, hardly affected by the numerical scale for the gradations of comparative judgement and rather insensitive to the choice of the cost-benefit relationship. Allocation procedures should also be fair. The benefits and/or the costs to be allocated to the parties in a distribution problem must be proportional to the effort, the strength, and/or the needs of the respective parties. The key observation in the present chapter is that proportionality can be pursued with methods which are based upon ratio information. The issues in question will be considered here via a number of case studies.

7.1 EVALUATION OF ENERGY RESEARCH PROGRAMS

The fundamental problem, when we allocate a research budget to competing research programs, is to find a compromise between conflicting criteria. The programs do not necessarily have the same impact under every criterion. Moreover, the relative importance of the criteria is vaguely known, at least at the start of the decision process. We are accordingly running up against an MCDA problem when we try to establish the benefits of the research programs, first by estimating the impacts and the criterion weights, and next by aggregating the results into final grades and/or terminal scores.

We do not merely face a problem of exclusive choice, however. The question is not to select the program with the highest benefits (the highest final grade or the highest terminal score) but to support each program at a level which is compatible with its benefits for a nation or a group of nations. The problem is even more complicated. Allocating financial resources to certain research programs would be meaningless if there are no competent research teams in the respective fields. Allocating financial resources to excellent proposals, however, might lead to an unbalanced distribution of research effort. In summary, the real problem is to support research areas where the research community is strong enough to deliver the expected results.

Finally, we cannot ignore the dimension of time. The allocation of a research budget is usually not an isolated event but part of a series of regular (mostly annual) budget adjustments. Thus, we may be confronted with a reallocation problem, and because research teams need several years to become fully productive, drastic changes in subsequent reallocations are mostly unfeasible.

We are now in a position to explain the scope of our case study: an exploration of the potential of MCDA in the reallocation of the European budget for non-nuclear energy research (Lootsma et al., 1990). The feasibility study commissioned by Directorate-General XII (Science, Research, and Technology) of the European Commission concentrated on the current research programs adopted for a period of four years (January 1985 – December 1988). It has been carried out in two phases: an initial phase with eight research programs and nine criteria in order to experiment with the methodology, and the definitive phase with ten programs and five criteria (December 1986 – June 1987). Six decision makers, members of the Directorate-General, were involved in the study during the whole period.

In the present chapter we summarize and discuss the results of MCDA in the second phase only, where we considered the following research programs with their current support levels in Millions of European Currency Units (1 ECU ≈ US$ 1.20 in late 1988).

1. Photovoltaic Cells. This program was particularly concerned with the design of solar power plants which, in the long term, might cover 7% of the European power production if the cell price could be reduced from 7 to 1 ECU/Watt. Current support level: 21.0 MECU.

2. Passive Solar Energy. This program was concerned with the improvement of techniques for heat transfer and for heat conservation in buildings. Current support level: 14.5 MECU.

3. Geothermal Energy. Objectives of the program: to extend the knowledge of available reserves of geothermal hot water and steam in the European Community, and to improve the techniques for winning or extraction. Current support level: 21.0 MECU.

4. Energy Saving. Objective: the development of advanced technologies for energy conservation, such as heat pumps for buildings and advanced batteries or fuel cells for electrical vehicles. Current support level: 8.0 MECU.

5. Combustion Processes in Industry. Objective: to improve the energy efficiency of the combustion processes in engines for the production of power and mechanical energy. Current support level: 18.5 MECU.

6. New Energy Vectors. Objective: the development of new and improved processes, both for the liquefaction of solid fossil fuels and for the production of coal-derived synthetic fuels. Current support level: 10.0 MECU.

7. Biomass Energy. Objective: to improve the conversion of biomass materials (waste, forestry, crops) into liquid and gaseous fuels. Current support level: 20.0 MECU.

8. Solid Fuels. Objectives: to improve the coal-combustion techniques via fluidized beds, to reduce the environmental effects of coal combustion, and to improve the methods for coal transportation. Current support level: 20.0 MECU.

9. Hydrocarbons. Objectives: to improve the methods for the discovery of oil deposits, to increase the production of oil from existing reservoirs, and to increase the use of natural gas. Current support level: 15.0 MECU.

10. Wind Energy. Objectives: to assess the wind energy potential of different regions in the European Community, to improve wind power generators, and to investigate grid connections. Current support level: 18.0 MECU.

Thus, the total four-years budget for the period 1985 – 1988 was 166.0 MECU, and the budget had to be reallocated at the end of 1988. In our feasibility study we asked the decision makers to judge whether the programs enhanced the following characteristics:

1. Security of Energy Supply. To which extent does the program contribute to the diversification of primary energy carriers?

2. Energy Efficiency. To which extent does the program increase the efficiency of energy production and use?

3. Long-Term Contribution. To which extent does the program contribute to the delayed depletion of primary energy carriers or to the consumption of renewable energy?

4. Environmental Protection. To which extent does the program support this issue?

5. Suitability of Community Action. To which extent does the program need Community support because there are significant economies of scale?

It is interesting to note that the following criteria had been dropped by the analysis in the initial phase: improvement of the competitive position of an innovation-based European industry, enhancement of the social acceptability of risky energy carriers (the nuclear issue), support of weakly developed regions in Europe, and contribution to the European independence policy in energy matters. These criteria attracted low weights in the experiments, and they hardly affected the terminal scores of the programs considered so far (Lootsma, 1988).

With the original data of 1987 (the pairwise comparisons of the ten programs under the five criteria just mentioned, carried out by the six members of DG XII), and with the methodology of the Chapters 2 - 4 we obtain the results displayed in Table 7.1. For the elicitation of the criterion weights, the gradations of relative importance are invariably encoded on a uniform scale with progression factor $\sqrt{2}$ so that the scale parameter γ has been set to the value 0.50. For the evaluation of the alternatives under the respective criteria, the gradations of relative performance have been converted into numerical values on three different geometric scales, see the corresponding values of the scale parameter γ. This mode of operation is a consequence of the analysis in Section 4.2, which also guarantees that the rank order of the terminal scores is preserved under variations of the scale parameter γ over a positive range of values. An example is given in Section 4.3, see also Table 4.7.

Table 7.1 *Weights of criteria for the allocation of an energy-research budget and terminal scores of alternative non-nuclear energy technologies judged on the basis of their possible contribution to the predominant goals of a European energy-research policy. The procedure since* 1992: *there is a uniform scale for the gradations of relative importance but the gradations of relative performance are converted into numerical values on three different geometric scales. Aggregation of ratio information via the calculation of geometric means.*

Criteria Potential to increase	Uniform scale for relative importance of criteria $\gamma = 0.50$			Rank Order
Security of Energy Supply		0.246		2
Energy Efficiency		0.154		4
Long-Term Contribution		0.256		1
Environmental Protection		0.239		3
Suitability for Community Action		0.106		5
Alternatives **Energy-research programs**	Varying scales for relative performance of alternatives $\gamma = 0.72$	$\gamma = 1.00$	$\gamma = 1.44$	Rank Order
Photovoltaic Cells	0.117	0.122	0.127	5
Passive Solar Energy	0.069	0.058	0.044	9
Geothermal Energy	0.087	0.080	0.070	7
Energy Saving	0.133	0.146	0.166	2
Combustion Processes	0.135	0.150	0.171	1
New Energy Vectors	0.077	0.069	0.056	8
Biomass Energy	0.056	0.043	0.029	10
Solid Fuels	0.119	0.125	0.132	3
Hydrocarbons	0.091	0.086	0.077	6
Wind Energy	0.117	0.122	0.127	4

In the AHP variant which we employed in 1987, however, we varied the gradations of relative importance and the gradations of relative performance simultaneously. Moreover, the aggregation into terminal scores was carried out via arithmetic-mean calculations although we were basically concerned with ratio information (see also our criticism on the original AHP in Section 3.6). Table 7.2 shows the results so obtained in the second phase of the project. We employed two scales (with $\gamma = 0.72$ and $\gamma = 1.44$) in order to prepare the sensitivity analysis of the proposed budget reallocation. Obviously, the rank order of the criterion weights has been preserved (this has mathematically been established in Section 4.2). There are several rank reversals in the terminal scores, however. That usually irritates the decision makers, although the terminal scores themselves are not dramatically affected by the scale variations.

In 1990 we systematically replaced the arithmetic-mean calculations by geometric-mean calculations in order to process ratio information. The advantages clearly emerged when we applied the new procedure to the original data of the project (Lootsma, 1990). The numerical values of the terminal scores did not undergo a dramatic shift, and the rank order appeared to be scale-independent. This is shown in Table 7.3. In general, however, the rank order of the terminal scores is only preserved under scale variations when we evaluate the criteria on a uniform scale (the key issue of Section 4.2).

Table 7.2 *Weights of criteria for the allocation of an energy-research budget and terminal scores of alternative non-nuclear energy technologies judged on the basis of their possible contribution to the predominant goals of a European energy-research policy. The procedure in the nineteen eighties: the gradations of relative importance and relative performance are simultaneously converted into numerical values on two different geometric scales. Aggregation of ratio information via the calculation of arithmetic means.*

Criteria	Varying scales for gradations of comparative judgement					
	$\gamma = 0.72$		$\gamma = 1.00$		$\gamma = 1.44$	
Potential to increase	Weights	Rank	Weights	Rank	Weights	Rank
Security of Energy Supply	0.261	2			0.294	2
Energy Efficiency	0.133	4			0.077	4
Long-Term Contrbution	0.277	1			0.332	1
Environmental Protection	0.251	3			0.272	3
Suitability for Community Action	0.078	5			0.026	5
Alternatives	$\gamma = 0.72$		$\gamma = 1.00$		$\gamma = 1.44$	
Energy-research programs	Scores	Rank	Scores	Rank	Scores	Rank
Photovoltaic Cells	0.117	5			0.136	4
Passive Solar Energy	0.071	9			0.052	9
Geothermal Energy	0.086	7			0.071	7
Energy Saving	0.130	2			0.144	2
Combustion Processes	0.134	1			0.168	1
New Energy Vectors	0.076	8			0.053	8
Biomass Energy	0.055	10			0.028	10
Solid Fuels	0.120	3			0.134	5
Hydrocarbons	0.091	6			0.077	6
Wind Energy	0.118	4			0.136	3

Table 7.3 *Weights of criteria for the allocation of an energy-research budget and terminal scores of alternative non-nuclear energy technologies judged on the basis of their possible contribution to the predominant goals of a European energy-research policy. The procedure in* 1990: *the gradations of relative importance and relative performance are simultaneously converted into numerical values on two or three different geometric scales. Aggregation of ratio information via the calculation of geometric means.*

Criteria	Varying scales for gradations of comparative judgement					
	$\gamma = 0.72$		$\gamma = 1.00$		$\gamma = 1.44$	
Potential to increase	Weights	Rank	Weights	Rank	Weights	Rank
Security of Energy Supply	0.261	2	0.277	2	0.294	2
Energy Efficiency	0.133	4	0.108	4	0.077	4
Long-Term Contrbution	0.277	1	0.302	1	0.332	1
Environmental Protection	0.251	3	0.262	3	0.272	3
Suitability for Community Action	0.078	5	0.051	5	0.026	5
Alternatives	$\gamma = 0.72$		$\gamma = 1.00$		$\gamma = 1.44$	
Energy-research programs	Scores	Rank	Scores	Rank	Scores	Rank
Photovoltaic Cells	0.118	3	0.126	3	0.137	3
Passive Solar Energy	0.069	9	0.060	9	0.047	9
Geothermal Energy	0.087	7	0.080	7	0.070	7
Energy Saving	0.132	2	0.141	2	0.154	2
Combustion Processes	0.136	1	0.153	1	0.180	1
New Energy Vectors	0.077	8	0.068	8	0.055	8
Biomass Energy	0.056	10	0.044	10	0.030	10
Solid Fuels	0.117	5	0.120	5	0.120	5
Hydrocarbons	0.091	6	0.085	6	0.076	6
Wind Energy	0.118	4	0.124	4	0.131	4

In the next two sections we shall be using the terminal scores of Table 7.2 in order to discuss the original results of the project in 1987. With the Multiplicative AHP and the terminal scores of Table 7.1, however, our conclusions would have been the same.

7.2 CONTINUOUS AND DISCRETE ALLOCATION MODELS

Obviously, we were not operating "in vacuo" but with on-going research programs. Drastic changes would heavily damage the current energy research. Hence, we considered only deviations from the current budget allocation. In its simplest form the reallocation problem can be written as the continuous knapsack problem of maximizing the objective function

$$b(c_1,\ldots,c_n) \tag{7.1}$$

subject to the constraints

$$\sum_{j=1}^{n} c_j \leq B, \tag{7.2}$$

$$l_j \leq c_j \leq u_j, j = 1,\ldots,n. \tag{7.3}$$

Here, c_j represents the support or cost level of the j-th research program, and B stands for the total budget of 166.0 MECU. The c_j may vary over the respective intervals (l_j, u_j) which are chosen so as to contain the current support levels

$$c_j^{*}.$$

The objective function b is the total benefit of the allocation $(c_1,\ldots, c_n)$. Usually, we simplify matters by assuming that b is a separable function so that it can be written as

$$\sum_{j=1}^{n} b_j(c_j).$$

The crucial problem now is to express the benefits of the individual research programs as functions of the support levels. In our study we started off by taking the terminal scores of the research programs (Table 7.2) to represent the benefits

$$b_j^{*} = b_j(c_j^{*})$$

at the current support levels. Note, however, that the terminal scores are not unique. In the pairwise-comparison steps, we only asked the comparative judgement of the decision makers, and we obtained ratio information only. For ease of exposition, the terminal scores are usually normalized, but in principle they have a multiplicative degree of freedom (see also Section 3.3). Thus, we merely assumed that the benefits at the current support levels are proportional to the terminal scores. This did not affect the optimal solutions to the allocation problems which we studied in the course of the project.

The results of a research project are rather sensitive to the financial resources allocated to it. A small reduction leads to cuts in travel and communication, thus isolating the research team from the scientific community. A small increase will enable the project leader to attract the qualified people and the sophisticated equipment, both necessary to obtain significant results. Hence, the benefits will increase monotonically with the support level, at least over the respective intervals (l_j, u_j). We finally decided to write

$$\frac{b_j(c_j)}{b_j^{*}} = \psi\left(\frac{c_j}{c_j^{*}}\right), \tag{7.4}$$

and to investigate the reallocation problem with four different types of cost-benefit relationships:

The convex relationship

$$\psi(\tau) = \tau^2. \tag{7.5}$$

The concave relationship

$$\psi(\tau) = \sqrt{\tau}. \tag{7.6}$$

The linear relationship

$$\psi(\tau) = \tau. \tag{7.7}$$

The S-shape relationship with the logistic function

$$\psi(\tau) = \frac{2}{1+\exp(4-4\tau)} \text{ so that } \psi'(\tau) = 2\psi(\tau)[2-\psi(\tau)]. \tag{7.8}$$

Each of these relationships represents a particular type of incremental returns to scale: increasing (7.5), decreasing (7.6), and constant (7.7), whereas (7.8) is typical for an activity with a slow take-off followed by rapid growth and saturation (see the behaviour of the first derivative). We did not have the illusion that the appropriate relationship for each of the respective research programs could easily be identified. Therefore, we decided to analyze the reallocation problem with each of the functions (7.5) – (7.8) in order to check whether the choice of a relationship would be critical.

So we had four variants of the objective function (7.1): a convex separable variant when (7.5) is substituted into (7.4), a concave separable variant when (7.6) is substituted into (7.4), etc. Although the differences seem to be marginal, the computational power of the appropriate solution algorithms varies widely. Maximization of a convex function over a polyhedron will generally produce a local maximum only; global optimality cannot be guaranteed. Maximization of a concave separable function over a polyhedron can rapidly and safely be carried out. Maximization of a linear function over a polyhedron is simply linear programming. Finally, maximization of a sum of S-shaped functions which are neither convex nor concave is hazardous.

Obviously, there is a sound computational argument to drop the continuous allocation model and to turn to a discrete model which can globally be solved by the current discrete optimization algorithms. This is not the only argument, however. It is a common experience in Operations Research that managers and administrators do not always think in terms of a continuum of alternatives. Frequently, they consider a small number of discrete alternatives only because each of them can easily be analyzed in detail. The alternative which is eventually selected does not always remain unaltered during the actual implementation. Nevertheless, the objectives of the foregoing selection process are usually attained: the preferences of the decision makers have been brought up and amply discussed. Hence, we formulated the problem of maximizing the total benefits as a discrete optimization problem with zero-one variables x_{jk} indicating whether the j-th research program should be supported at the k-th cost level, yes or no. In this form the reallocation problem is known as a research portfolio selection problem. Taking b_{jk} and c_{jk} to denote the associated benefits and costs, we maximized the total benefits

$$\sum_j \sum_k b_{jk} x_{jk} \tag{7.9}$$

subject to the budget constraint

$$\sum_j \sum_k c_{jk} x_{jk} \leq B \quad (166.0 \text{ MECU}) \tag{7.10}$$

and to the logical constraints

$$\sum_k x_{jk} = 1 \text{ for every project } j, \tag{7.11}$$

stipulating that each research program must be executed at exactly one of the possible cost levels. In consultation with the members of Directorate-General XII we introduced for each research program five possible cost or support levels: the current level, and cost levels at 60%, 80%, 120%, and 140% of the current level. In principle we could have introduced the zero level as well, so that there would have been an option to drop certain energy technologies completely from the program. We were again confronted with four variants of the objective function by setting

$$\frac{b_{jk}}{b_j^*} = \psi\left(\frac{c_{jk}}{c_j^*}\right),$$

where b_j^* and c_j^* denoted the current benefits and costs, whereas

$$\frac{c_{jk}}{c_j^*} = 0.60,\ 0.80.\ 1.00,\ 1.20,\ 1.40.$$

For the function ψ we subsequently used each of the forms (7.5) – (7.8). To illustrate matters, let us consider the Photovoltaic Cells program (current 100% cost level 21.0 MECU) which had the terminal score 11.7 (the 100% benefits) on the scale with $\gamma = 0.72$ (see Table 7.2). Table 7.4 shows the five possible cost levels and their associated benefits if we adopt the convex quadratic cost-benefit relationship (7.5).

It will be clear that the benefits are scale-dependent so that we do not only have to analyze the consequences of the presumed cost-benefit relationship but also the effect of scale variations.

Table 7.4 *Selected cost levels and their associated benefits under a convex quadratic cost-benefit relationship.*

Costs (MECU)		Benefits		Benefits/MECU
60%	12.6	36%	4.2	0.33
80%	16.8	64%	7.5	0.45
100%	21.0	100%	11.7	0.56
120%	25.2	144%	16.8	0.67
140%	29.4	196%	22.9	0.78

7.3 OPTIMIZATION AND SENSITIVITY ANALYSIS

The solution of the reallocation problem (7.9) – (7.11), with the convex quadratic cost-benefit relationship $\psi(\tau) = \tau^2$ and with the current benefits set to the terminal scores on the scale with $\gamma = 0.72$ (see Table 7.2), may be found bold-faced in Table 7.5. It will be obvious that particularly the programs with high benefits/MECU are suitable for support at higher cost levels. Table 7.6 shows a reallocation in bold type when the benefits are derived from the terminal scores on the scale with $\gamma = 1.44$ (see also Table 7.2). Again, there is a tendency to increase the financial support for programs with high benefits/MECU.

The rank order of the research programs on the basis of the benefits/MECU appeared to be rather scale-dependent, as Table 7.7 readily shows. Nevertheless, there was a consistent pattern in the reallocation. On both scales the programs with high (low) benefits/MECU were recommended for support at the highest (lowest) level, but there was a middle group of programs (Photovoltaic Cells, New Energy Vectors, and Hydrocarbons) at strongly alternating support levels.

This pattern also emerged when we varied the total budget B in the neighbourhood of 166.0 MECU. Using the terminal scores on the scale with $\gamma = 0.72$, we started off by setting B to the non-integer value 180.05, and we obtained an optimal solution where the sum of the reallocated amounts was 179.9 MECU. We eliminated this solution by setting $B = 179.85$, whereafter we restarted the optimization process, etc. We applied the same approach with the terminal scores taken from the scale with $\gamma = 1.44$. The reader may find the results of these budget variations in Lootsma et al. (1990).

Of course, we also analyzed the reallocation problem (7.9) – (7.11) with the concave cost-benefit relationship (7.6), the linear relationship (7.7), and the S-shape relationship (7.8). Remarkably enough, the solutions were almost the same as with the convex relationship (7.5). On the basis of this elaborate sensitivity analysis we could eventually summarize our conclusions as shown in Table 7.7.

Table 7.5 *Benefits of non-nuclear energy-research programs at the current support level (middle column,* 100%) *and at predetermined lower* (60%, 80%) *and higher* (120%, 140%) *support levels. The benefits at the current level are the terminal scores* (× 100) *of the programs on the scale with* $\gamma = 0.72$ *in Table* 7.2. *The benefits at the other levels are derived from a hypothetical convex quadratic cost-benefit relationship. The optimal solution of the discrete reallocation problem is bold-faced. Total costs* 165.2 MECU, *total benefits* 134.3 *units.*

		60%	80%	100%	120%	140%
	Benefits	**4.2**	7.5	11.7	16.8	22.9
Photovoltaic Cells	Costs	**12.6**	16.8	21.0	25.2	29.4
	Benefits/MECU	**0.33**	0.45	0.56	0.67	0.78
	Benefits	**2.6**	4.5	7.1	10.2	13.9
Passive Solar En.	Costs	**8.7**	11.6	14.5	17.4	20.3
	Benefits/MECU	**0.29**	0.39	0.49	0.59	0.69
	Benefits	**3.1**	5.5	8.6	12.4	16.9
Geothermal En.	Costs	**12.6**	16.8	21.0	25.2	29.4
	Benefits/MECU	**0.25**	0.33	0.41	0.49	0.57
	Benefits	4.7	8.3	13.0	18.7	**25.5**
Energy Saving	Costs	4.8	6.4	8.0	9.6	**11.2**
	Benefits/MECU	0.98	1.30	1.63	1.95	**2.27**
	Benefits	4.8	8.6	13.4	19.3	**26.3**
Combustion Pr.	Costs	11.1	14.8	18.5	22.2	**25.9**
	Benefits/MECU	0.43	0.58	0.72	0.87	**1.01**
	Benefits	2.7	4.9	7.6	10.9	**14.9**
New En. Vectors	Costs	6.0	8.0	10.0	12.0	**14.0**
	Benefits/MECU	0.46	0.61	0.76	0.91	**1.06**
	Benefits	**2.0**	3.5	5.5	7.9	10.8
Biomass Energy	Costs	**12.0**	16.0	20.0	24.0	28.0
	Benefits/MECU	**0.17**	0.22	0.28	0.33	0.39
	Benefits	4.3	7.7	12.0	17.3	**23.5**
Solid Fuels	Costs	12.0	16.0	20.0	24.0	**28.0**
	Benefits/MECU	0.36	0.48	0.60	0.72	**0.84**
	Benefits	3.3	5.8	**9.1**	13.1	17.8
Hydrocarbons	Costs	9.0	12.0	**15.0**	18.0	21.0
	Benefits/MECU	0.36	0.49	**0.61**	0.73	0.85
	Benefits	4.2	7.6	11.8	17.0	**23.1**
Wind Energy	Costs	10.8	14.4	18.0	21.6	**25.2**
	Benefits/MECU	0.39	0.52	0.66	0.79	**0.92**

Table 7.6 *Benefits of non-nuclear energy-research programs at the current support level (middle column,* 100%) *and at predetermined lower* (60%, 80%) *and higher* (120%, 140%) *support levels. The benefits at the current level are the terminal scores* (× 100) *of the programs on the scale with* $\gamma = 1.44$ *in Table* 7.2. *The benefits at the other levels are derived from a hypothetical convex quadratic cost-benefit relationship. The optimal solution of the discrete reallocation problem is bold-faced. Total costs* 166.0 MECU, *total benefits* 145.5 *units.*

		60%	80%	100%	120%	140%
	Benefits	4.9	8.7	13.6	19.6	**26.7**
Photovoltaic Cells	Costs	12.6	16.8	21.0	25.2	**29.4**
	Benefits/MECU	0.39	0.52	0.65	0.78	**0.91**
	Benefits	**1.9**	3.3	5.2	7.5	10.2
Passive Solar En.	Costs	**8.7**	11.6	14.5	17.4	20.3
	Benefits/MECU	**0.22**	0.29	0.36	0.43	0.50
	Benefits	**2.6**	4.5	7.1	10.2	13.9
Geothermal En.	Costs	**12.6**	16.8	21.0	25.2	29.4
	Benefits/MECU	**0.20**	0.27	0.34	0.41	0.47
	Benefits	5.2	9.2	14.4	20.7	**28.2**
Energy Saving	Costs	4.8	6.4	8.0	9.6	**11.2**
	Benefits/MECU	1.08	1.44	1.80	2.16	**2.52**
	Benefits	6.0	10.8	16.8	24.2	**32.9**
Combustion Pr.	Costs	11.1	14.8	18.5	22.2	**25.9**
	Benefits/MECU	0.54	0.73	0.91	1.09	**1.27**
	Benefits	1.9	**3.4**	5.3	7.6	10.4
New En. Vectors	Costs	6.0	**8.0**	10.0	12.0	14.0
	Benefits/MECU	0.32	**0.42**	0.53	0.64	0.74
	Benefits	**1.0**	1.8	2.8	4.0	5.5
Biomass Energy	Costs	**12.0**	16.0	20.0	24.0	28.0
	Benefits/MECU	**0.08**	0.11	0.14	0.17	0.20
	Benefits	4.8	8.6	13.4	**19.3**	26.3
Solid Fuels	Costs	12.0	16.0	20.0	**24.0**	28.0
	Benefits/MECU	0.40	0.54	0.67	**0.80**	0.94
	Benefits	**2.8**	4.9	7.7	11.1	15.1
Hydrocarbons	Costs	**9.0**	12.0	15.0	18.0	21.0
	Benefits/MECU	**0.31**	0.41	0.51	0.62	0.72
	Benefits	4.9	8.7	13.6	19.6	**26.7**
Wind Energy	Costs	10.8	14.4	18.0	21.6	**25.2**
	Benefits/MECU	0.45	0.60	0.76	0.91	**1.06**

Table 7.7 *Non-nuclear energy-research programs with their benefits/MECU at the current and at the optimum support levels. Recommended action when the total budget varies over the range* 160 – 180 MECU: *increase the financial support, possibly to the* 140% *level, decrease the support, possibly to the* 60% *level, or maintain the support roughly at the* 100% *level.*

Benefits/MECU	at current level		at optimum level				Recommendation
	$\gamma = 0.72$	$\gamma = 1.44$	$\gamma = 0.72$		$\gamma = 1.44$		for fin. support
Photovoltaic Cells	0.56	0.65	60%	0.33	140%	0.91	maintain
Passive Solar En.	0.49	0.36	60%	0.29	60%	0.22	decrease
Geothermal En.	0.41	0.34	60%	0.25	60%	0.20	decrease
Energy Saving	1.63	1.80	140%	2.27	140%	2.52	increase
Combustion Pr.	0.72	0.91	140%	1.01	140%	1.27	increase
New En. Vectors	0.76	0.53	140%	1.06	80%	0.42	maintain
Biomass Energy	0.28	0.14	60%	0.17	60%	0.08	decrease
Solid Fuels	0.60	0.67	140%	0.84	120%	0.80	increase
Hydrocarbons	0.61	0.51	100%	0.61	60%	0.31	maintain
Wind Energy	0.66	0.76	140%	0.92	140%	1.06	increase

In summary, there were two major uncertainties in the reallocation problem. First, there was no unique scale to quantify verbal human judgement. Hence, we calculated the terminal scores (the benefits) of the research programs via two different scales. Second, the relationship between the benefits and the costs of a research program was only vaguely known (presumably increasing in the neighbourhood of the current support level). This prompted us to calculate optimal reallocations for various types of relationships. In addition, we varied the total budget in order to overcome the notorious instability of solutions to discrete optimization problems. There was a clear and persistent pattern in the proposed reallocations. The predominant factor to decide whether one could decrease or increase the support level of a research program or whether one should leave it unchanged was given by the benefits/MECU.

In our feasibility study we only asked the decision makers to judge the existing allocation and we studied several possible reallocations on the basis of hypothetical cost-benefit relationships. An alternative mode of operation would have been to submit various hypothetical and plausible reallocations to the decision makers and to ask for their comparative pairwise judgement. In the latter case the amount of work might have been prohibitive, however.

Even when we finished a manuscript for possible publication in a scientific journal (early 1989), definitive reallocation decisions were still outstanding. There were rumours that the total budget for non-nuclear energy research would be reduced from 166 to 123 MECU. Such a budget cut was deeper than we originally expected. Technically speaking, however, we could immediately come up with new recommendations. The only thing to do was to continue the sensitivity analysis and to calculate optimal reallocations for total budget levels varying between 110 and 130 MECU. Although this is not really a marginal variation in the neighbourhood of 166 MECU, the results were sufficiently interesting to be summarized. The analysis suggested that Energy Saving should be supported at the 140% level, despite the budget cut, and that the support for Combustion Processes should at least be maintained at the current 100% level. Financial support for the remaining programs should be reduced to the 60% level.

Because many decisions in international organisations are made by representatives of countries varying considerably in size and power, we extended the MCDA method which we employed so that power coefficients could be assigned to the decision makers themselves. A detailed description of the resulting method for group decision making may be found in Chapter 6. The terminal scores so obtained would hopefully lead to an acceptable compromise in international decision making. The feasibility study has therefore not been followed by a full-fledged study (see also Section 6.3). When we eventually asked the Directorate-General XII for permission to publish the methodological results of the project we immediately received a negative answer. *"The publication could be harmful for the existing good political climate between the policy makers of the European Community and the member states. The European Commission does not want to circulate the results of the analysis of a single program (non-nuclear energy research) based upon the value judgements of internal experts only".* The research contracts with our university, however, stipulate that the publication of methodological results cannot be delayed by more than two years. That brought an end to the dispute.

7.4 FAIRNESS AND EQUITY

In the sections to follow we are concerned with principles of fairness and equity in order to incorporate them in a method for the allocation of benefits or costs (the output) in a distribution problem, on the basis of the effort, the strength or the needs (the input) of the respective parties. This is an old question. Even in the Antiquity distributive justice was a point of discussion. "*All men agree that what is just in distribution must be according to merit in some sense, but they do not specify the same sort of merit*" (Aristotle's Ethic, in the translation by J. Warrington, 1963). This statement briefly summarizes the two issues to be discussed here. First, the leading principle in fairness and equity is proportionality, which means that the benefits and the costs (the output) to be allocated to the parties in a distribution problem must be proportional to the effort, the strength, and/or the needs (the input) of the respective parties. Second, since input and output are usually measured under several criteria so that they are multi-dimensional, we have to weigh the distribution criteria in order to establish the aggregate quantities that must be proportional.

There are obvious utilitarian reasons why proportionality should be the leading principle of fairness. People are mostly unwilling to contribute relatively high inputs unless they can look forward to relatively high outputs. Moreover, a person who can more effectively use. It is not unusual, however, that people moderate or amplify a proportional distribution. Young (1994) proposes four principles to determine a claimant's share: proportionality, progressivity, parity, and priority. The rationale for the progressivity principle, usually found in taxation schemes, is that those who are better off should pay at a higher rate because they can absorb the loss more easily. Under the egalitarian parity principle benefits and costs are allocated equally, even if the parties are unequal. The priority principle is an "all or nothing" principle which assigns absolute precedence to one party in the allocation of benefits and costs. It is usually applied to distribute indivisible goods.

Fishburn and Sarin (1994) discuss the issues of fairness and equity in the broad perspective of social choice. Fairness is based upon the preferences of individuals and groups, and upon the ways in which they perceive themselves in relation to others. Distributions that are fairest are those in which there is little or no envy among parties. Equity is based upon external ethical criteria, not on the specific preferences of individuals or groups. It is usually interpreted as some sort of equality, meaning that people are morally equal and should be treated with equal concern and respect. A distinction can also be made between fairness of outcome and procedural fairness (Linnerooth-Bayer et. al., 1994). In the first case the emphasis is on the results of the distribution process (do the parties agree with the proposed shares?), in the second case on the distribution process itself and on the role of the respective parties in it (did they receive a fair treatment, did they have a fair opportunity to explain their viewpoints?).

We shall not further discuss here the extensive literature on the principles of fairness and equity and on distributive justice. Note that the literature is sometimes confusing. Many highly similar concepts appear under different names. It is not always clear how they could be made operational (that is in fact the objective of the present section). For more information we refer the reader to Deutsch (1975, 1985), Kasperson (1983), and Messick and Cook (1983).

By the principle of proportionality we are led to the algorithmic steps of the Multiplicative AHP because this method is particularly designed for the elicitation and the processing of ratio information. Thus, we will use logarithmic regression in order to analyze desired-ratio matrices, and we will apply geometric-mean calculations in order to work with a variety of distribution criteria (Lootsma et al., 1998). Moreover, the method can easily be used under the other principles of fairness as well. All one has to do is to employ powers of the desired ratios. With exponents greater (smaller) than one we model the principle of progressivity (moderation). In the limiting cases, when the exponents tend to infinity (zero), we proceed to the priority (parity) principle. Let us now try to model the distribution of resources under these principles.

We consider a distribution problem with m criteria and n parties, first under the principle of proportionality. We take the symbol r_{ijk} to represent the desired ratio of the contributions c_j and c_k to be made by the respective parties under criterion i. Let us further introduce the symbol $R_i = \{r_{ijk}\}$ to stand for the matrix of the desired ratios under the i-th criterion. This matrix is positive and reciprocal, but not necessarily consistent, just like a pairwise-comparison matrix in the AHP. It may happen that $r_{ijk} \times r_{ikl} \neq r_{ijl}$ (for an example, see Section 7.6). Let w_i stand for the weight assigned to the i-th distribution criterion. Following the mode of operation in the Multiplicative AHP we take the ratio c_j/c_k of any pair of contributions to approximate the desired ratios $r_{1jk},\ldots,r_{mjk}$ simultaneously, in the sense that we solve the contributions from the logarithmic-regression problem of minimizing

$$\sum_{i=1}^{m}\sum_{j>k} w_i\left({}^2\log r_{ijk} - {}^2\log c_j + {}^2\log c_k\right)^2. \tag{7.12}$$

Actually, we carry out the unconstrained minimization by solving the associated linear system of normal equations with the variables $u_j = \ln c_j$, $j = 1,\ldots, n$. Obviously, the u_j have an additive degree of freedom. The c_j will accordingly have a multiplicative degree of freedom. A particular solution to the regression problem is given by

$$c_j = \prod_{i=1}^{m}\left(\sqrt[n]{\prod_{l=1}^{n} r_{ijl}}\right)^{w_i}, \tag{7.13}$$

which can be obtained if we calculate first the geometric row means of the matrices R_i and thereafter the geometric means of the row means. These operations may be interchanged without altering the final results. If the desired-ratio matrices happen to be consistent, the ratio of any pair of contributions is uniquely given by

$$\frac{c_j}{c_k} = \prod_{i=1}^{m}\left(\sqrt[n]{\prod_{l=1}^{n}\frac{r_{ijl}}{r_{ikl}}}\right)^{w_i} = \prod_{i=1}^{m}\left(r_{ijk}\right)^{w_i}. \tag{7.14}$$

Let us illustrate the above results via the allocation of fair contributions to the European Union, to be paid annually by the member states. If the respective contributions must be proportional to the size of the population and the Gross Domestic Product, we have two diverging requirements that can only approximately be satisfied. Suppose that equal weights are assigned to these distribution criteria. We take the ratio c_j/c_k of any pair of contributions to approximate the desired ratios

$$r_{1jk} = \frac{Pop_j}{Pop_k}$$

and

$$r_{2jk} = \frac{GDP_j}{GDP_k}$$

simultaneously by the solution of the above logarithmic regression problem. On the basis of formula (7.14) the ratio c_j/c_k can now be written as

$$\frac{c_j}{c_k} = \sqrt{\frac{Pop_j}{Pop_k} \times \frac{GDP_j}{GDP_k}}\,. \tag{7.15}$$

If we also want to the use the national area as a yardstick to set the contributions, we introduce the desired ratios

$$r_{3jk} = \frac{Area_j}{Area_k}.$$

By formula (7.14) the ratio of any pair of contributions is now given by

$$\frac{c_j}{c_k} = \sqrt[3]{\frac{Pop_j}{Pop_k} \times \frac{GDP_j}{GDP_k} \times \frac{Area_j}{Area_k}}\,, \tag{7.16}$$

at least if equal weights are assigned to the distribution criteria. The choice of the criterion weights in general is still under investigation. It is unclear, for instance, how large the weights should be in order to represent various gradations of relative importance of the distribution criteria. A detailed discussion of fair contributions to the European Union may be found in Section 7.5.

A refinement of the model is to replace the r_{ijk} by powers $(r_{ijk})^{q_i}$. This will enable us to work under the remaining principles of fairness. The positive exponent q_i introduces a moderation (amplification) of the desired ratios under the i-th distribution criterion if $q_i <$ 1 ($q_i > 1$). By the introduction of this exponent we model the principle of progressivity. For very small (very large) values of q_i there is a transition from the principle of proportionality to the principle of parity (priority). An application of the idea is presented in Section 7.6, where we are concerned with a fair seat allocation in the European Parliament.

7.5 FAIR CONTRIBUTIONS TO THE EUROPEAN UNION

The data to be employed here may be found in the Eurostat Yearbook (1995). They show the state of the European Union in 1993. Table 7.8 exhibits the size of the population and the Gross Domestic Product of the respective member countries in that year, as well as the national area.

Table 7.8 *Population, Gross Domestic Product, and national area of the* 12 *member states of the European Union in* 1993. *Data from the Eurostat Yearbook* 1995, *pages* 72, 168, *and* 196.

	Population 1993 in millions	GDP 1993 in 10^9 ECU	National Area in 1000 sq km
Belgium	10.07	125	31
Denmark	5.18	85	43
Germany	80.98	1105	357
Greece	10.35	59	132
Spain	39.05	277	505
France	57.53	810	549
Ireland	3.56	37	70
Italy	56.96	658	301
Luxembourg	0.40	8	3
The Netherlands	15.24	205	41
Portugal	9.87	41	92
United Kingdom	58.10	709	244
European Union	347.29	4119	2368

Table 7.9 *Actual contributions of the* 12 *member states to the European Union. Data from the Eurostat Yearbook* 1995, *page* 402.

	Contribution, 10^9 ECU	Percentage of GDP	ECU per capita
Belgium	2.39	1.91	237
Denmark	1.21	1.42	233
Germany	19.03	1.72	235
Greece	1.01	1.72	98
Spain	5.19	1.88	133
France	11.56	1.43	201
Ireland	0.57	1.53	159
Italy	10.08	1.53	177
Luxembourg	0.17	2.10	420
The Netherlands	4.02	1.96	264
Portugal	0.91	2.21	92
United Kingdom	7.61	1.07	131
European Union	63.75	1.55	184

The resources of the European Union (63.75 billion ECU) consisted of customs revenues (16.8%), levies on agricultural imports and sugar storage (2.9%), a VAT-based levy (52.2%), a GDP-based contribution (25.2%), and non-attributable income (2.6%). The total contribution of each of the member countries in 1993 is exhibited in Table 7.9, which also shows how the contributions are related to the national economies. The contributions expressed as a percentage of the GDP vary between 1.07 (United Kingdom) and 2.21 (Portugal), the contributions in ECU per capita between 92 (Portugal) and 420 (Luxembourg). Since the ratio of the smallest to the largest contribution per capita is roughly 1:5, fairness is far to seek.

Table 7.10 *Possible contributions of the member states to the European Union on the basis of the size of the population.*

	Contribution, 10^9 ECU	Percentage of GDP	ECU per capita
Belgium	1.84	1.47	184
Denmark	0.95	1.11	184
Germany	14.87	1.34	184
Greece	1.90	3.22	184
Spain	7.16	2.58	184
France	10.56	1.30	184
Ireland	0.66	1.78	184
Italy	10.45	1.59	184
Luxembourg	0.08	1.00	184
The Netherlands	2.80	1.37	184
Portugal	1.81	4.41	184
United Kingdom	10.67	1.50	184
European Union	63.75	1.55	184

Table 7.11 *Possible contributions of the member states to the European Union on the basis of the Gross Domestic Product.*

	Contribution, 10^9 ECU	Percentage of GDP	ECU per capita
Belgium	1.93	1.55	192
Denmark	1.31	1.55	253
Germany	17.10	1.55	211
Greece	0.91	1.55	88
Spain	4.28	1.55	110
France	12.53	1.55	218
Ireland	0.57	1.55	160
Italy	10.18	1.55	179
Luxembourg	0.12	1.55	300
The Netherlands	3.17	1.55	208
Portugal	0.63	1.55	64
United Kingdom	10.97	1.55	189
European Union	63.75	1.55	184

The Tables 7.10 and 7.11 show that contributions proportional to the size of the population or the GDP only (one single distribution criterion) are also unfair. When only the size of the population is used, there is indeed a uniform contribution of 184 ECU per capita, but the contributions expressed as a percentage of the GDP vary between 1.00 (Luxembourg) and 4.41 (Portugal). Similarly, with the GDP as the unique yardstick, the contributions are uniformly set to 1.55% of the GDP, but the contributions in ECU per capita vary between 64 (Portugal) and 300 (Luxembourg). So, the ratio of the smallest to the largest contribution per capita is roughly 1:5, although the GDP is generally considered to be a proper yardstick for a country's ability to pay (see also Beckermann (1980)).

Table 7.12 *Possible contributions of the member states to the European Union on the basis of the geometric mean of the population and the Gross Domestic Product.*

	Contribution, 10^9 ECU	Percentage of GDP	ECU per capita
Belgium	1.91	1.53	190
Denmark	1.13	1.32	218
Germany	16.07	1.45	198
Greece	1.32	2.24	128
Spain	5.58	2.01	143
France	11.60	1.43	202
Ireland	0.62	1.68	174
Italy	10.40	1.58	183
Luxembourg	0.10	1.25	250
The Netherlands	3.01	1.47	198
Portugal	1.08	2.63	109
United Kingdom	10.91	1.54	188
European Union	63.75	1.55	184

Table 7.13 *Possible contributions of the member states to the European Union on the basis of the geometric mean of the population, the Gross Domestic Product, and the national area.*

	Contribution, 10^9 ECU	Percentage of GDP	ECU per capita
Belgium	1.50	1.20	149
Denmark	1.18	1.39	228
Germany	14.07	1.27	174
Greece	1.92	3.25	186
Spain	7.80	2.82	200
France	13.06	1.61	227
Ireland	0.94	2.54	264
Italy	9.94	1.51	175
Luxembourg	0.10	1.25	250
The Netherlands	2.22	1.08	146
Portugal	1.48	3.61	150
United Kingdom	9.56	1.35	165
European Union	63.75	1.55	184

A considerable improvement is obtained when we allocate the contributions on the basis of the size of the population and the GDP simultaneously, with equal weights so that formula (7.15) applies. Table 7.12 shows that the contributions expressed as a percentage of the GDP now vary between 1.25 (Luxembourg) and 2.63 (Portugal), whereas the contributions in ECU per capita vary between 109 (Portugal) and 250 (Luxembourg). Under both distribution criteria the ratio of the smallest to the largest contribution is reduced to 1:2. The third distribution criterion that may come up in the discussions on fairness and equity is the national area. A large area has many possible advantages for a country: a large amount of arable land and fresh water to support agriculture, large mountainous regions to support tourism and water winning, large spaces to enhance the quality of life, and/or large mineral deposits or fossil-fuel supplies. Table 7.13 shows the possible contributions to the European Union when the size of the population, the GDP, and the national area are used with equal weights to distribute the total burden. Formula (7.16) is clearly applicable. The contributions now vary between 1.08% (The Netherlands) and 3.61% (Portugal) of the GDP, and between 146 ECU (The Netherlands) and 264 ECU (Ireland) per capita. The ratio of the smallest to the largest percentage of the GDP is higher than 1:2 now, but this may have a good reason: the population densities in the European Union vary widely, between 51 inhabitants per km^2 (Ireland) and 370 (The Netherlands). The third distribution criterion is clearly an incentive for the member states to exploit their natural resources more effectively.

Note. In the present paper we ignored the attempts of various European countries to use the principle of "juste retour" in order to regain their contributions to the Union as much as possible.

7.6 SEAT ALLOCATIONS IN THE EUROPEAN PARLIAMENT

With some moderation the principle of proportionality can also be used to explain the seat allocation in the European Parliament as a function of the size of the population of the member states. Because we have only one distribution criterion here we can simplify formula (7.12). We carry out the study via unconstrained and constrained minimization of the sum of squares

$$\sum_{j>k}\left({}^2\log r_{jk} - {}^2\log s_j + {}^2\log s_k\right)^2, \tag{7.17}$$

where s_j and s_k denote the size of the delegations of the respective member states j and k. Let us first introduce new symbols to represent the actual population ratios

$$\alpha_{jk} \quad = \quad \frac{Pop_j}{Pop_k}. \tag{7.18}$$

Table 7.14 shows the respective seat allocations obtained by unconstrained minimization of the sum of squares (7.17) when the desired ratio of any pair of national delegations is taken to be

$$r_{jk} = \alpha_{jk}^q, \quad q = 1, 0.9, 0.8, 0.7.$$

The non-integer solution of the minimization process must be adjusted to an integer seat allocation via an apportionment rule (Webster's rule, for instance, see Young, 1994).

Table 7.14 *Number of seats in the European Parliament, proportional to the size of the population of the member states (column* 2)*, moderately proportional to the size of the population (columns* 3 - 5)*, as well as the actual allocation of seats in* 1993 *(column* 6)*. Data from the Eurostat Yearbook* 1995.

	Pop	Pop^{09}	Pop^{08}	Pop^{07}	Actual
Belgium	16	19	22	25	25
Denmark	8	10	13	16	16
Germany	132	124	116	108	99
Greece	17	20	22	26	25
Spain	64	65	65	65	64
France	94	91	88	85	87
Ireland	6	7	10	12	15
Italy	93	91	88	84	87
Luxembourg	1	1	2	3	6
The Netherlands	25	28	30	33	31
Portugal	16	19	22	25	25
United Kingdom	95	92	89	85	87
Eur. Parliament	567	567	567	567	567

Table 7.15 *Number of seats in the European Parliament, proportional to the size of the population (column* 2)*, calculated with truncated population ratios (column* 3) *and with constrained population ratios (column* 4)*, as well as the actual allocation of seats in* 1993 *(column* 5)*. Data from the Eurostat Yearbook* 1995.

	Pop	Pop, truncated	Pop, constrained	Actual
Belgium	16	19	21	25
Denmark	8	10	11	16
Germany	132	122	96	99
Greece	17	19	21	25
Spain	64	65	69	64
France	94	92	95	87
Ireland	6	7	8	15
Italy	93	92	94	87
Luxembourg	1	2	6	6
The Netherlands	25	27	30	31
Portugal	16	19	20	25
United Kingdom	95	93	96	87
Eur. Parliament	567	567	567	567

With $q = 1$, we have pure proportionality on the basis of the size of the population. Smaller values of q lead to more equality between the delegations. Remarkably enough, the seat allocation for $q = 0.7$ practically coincides with the actual seat allocation in the European Parliament. At the extreme ends of the spectrum, however, we find some discrepancies. There are two countries with an over-representation, Ireland and Luxembourg, and one with an under-representation, Germany. An issue to be put on the political agenda?

Let us now use the seat-allocation problem in order to illustrate the introduction of certain constraints in a distribution problem. In Section 6.3, considering the power relations in groups, we proposed to limit the ratio of any two delegations to the range between $^1/16$ and 16 in order to prevent the total domination of the stronger parties over the weaker ones. Column 6 in Table 7.14 shows that these constraints are actually satisfied in the European Parliament despite the discrepancies between Luxembourg and Germany. We have also tried to achieve this via unconstrained and via constrained minimization of the sum of squares (7.17). In the first approach we truncated the desired ratios so that

$$r_{jk} = \begin{cases} 1/16, & \alpha_{jk} < 1/16, \\ 16, & \alpha_{jk} > 16, \\ \alpha_{jk}, & 1/16 \le \alpha_{jk} \le 16. \end{cases}$$

The desired-ratio matrix with these elements is inconsistent. Unconstrained minimization of the sum of squares (7.17) provides a non-integer solution so that one needs again an apportionment rule to obtain an integer seat allocation. The result is shown in column 3 of Table 7.15. Some ratios are still beyond the desired range, however, see the size of the delegations of Luxembourg (2 seats) and Germany (122 seats). In the second approach we minimize the sum of squares (7.17) with $r_{jk} = \alpha_{jk}$ under the constraints

$$-\,{}^2\log 16 \le {}^2\log s_j - {}^2\log s_k \le {}^2\log 16, \tag{7.19}$$

for any j and k. This is a convex quadratic-programming problem when the logarithms of the delegations stand for the variables. In the non-integer optimal solution, Luxembourg appears to have a delegation of 5.5 seats whereas the large countries France, Germany, Italy, and the UK have 88 seats each. An apportionment rule will not change the size of the delegations of the large countries, but if the delegation of Luxembourg is reduced to 5, some constraints in (7.19) will be violated, and if the delegation of Luxembourg is rounded off to 6, the constraints (7.19) will be inactive. Column 4 of Table 7.15 shows the result when the delegation of Luxembourg has been set to 6 before the start of the constrained minimization process. The ratio of the delegations of Luxembourg (6 seats) and Germany (96 seats) is now precisely within the required range.

7.7 OUTPUT-BASED RESEARCH FUNDING

The idea to use desired ratios in allocation problems can successfully be applied in other areas as well. We consider here the evaluation of academic research, and particularly the assignment of scores to research output items such as articles, books, contributions to conference proceedings, doctoral dissertations, etc., under four distribution criteria: the advancement of pure and applied scientific research, the advancement of innovative problem solving, the dissemination of scientific knowledge, and the coverage of the cost of the research. The resulting scores are to be used in the allocation of research funds to the faculties of a university. This is a delicate issue.

Since 1995 the Delft University of Technology allocates research funds to the faculties on the basis of the annual research output: the number of articles, books, doctoral dissertations, etc., in a given year. A score is assigned to each output item: articles in outstanding scientific journals 5, books published by outstanding scientific publishing companies 10, doctoral dissertations 10, contributions to conference proceedings, if they are refereed 3, otherwise 2, editorship of conference proceedings 2, etc.. On the basis of these scores the total university budget for research, the so-called first stream of financial resources, is proportionally allocated to the faculties. For various administrative reasons there is a time lag. The budget for the year n is allocated in the year $n-1$ on the basis of the output in the year $n-2$ (and hence on the basis of ideas in the years $n-4$ or $n-5$). The Erasmus University Rotterdam has adopted a similar system for the allocation of the research budget. For a description of the underlying bibliometric methods we refer the reader to Van Raan (1993, 1996). Such a system raises several questions. Does it really make sense? What are the objectives? Does the system promote pure scientific research, is it an incentive for multi-disciplinary and/or applied research, does it properly cover the effort of the scientific staff members and the expenses of the faculties, and/or does it promote undesirable strategic behaviour? Does the system eventually lead to an improved reputation of the university? Does it leave the scientists sufficient time for services to the scientific community (referee reports) and to the society in general? And if the objectives lead to contradictory answers, what is the relative importance of the corresponding distribution criteria?

Given the steadily increasing cost of higher education and the need for administrators to control the expenditures, output-based research funding is likely to be implemented in several universities, despite the administrative overheads. Since it was our objective to contribute to the overall aims of the Decision Analysis and Support (DAS) project of the International Institute for Applied Systems Analysis (IIASA, Laxenburg, Austria) via realistic experiments, we submitted the following case study to the participants of the DAS workshop, Delft, November 1997 (see Lootsma and Bots, 1999).

The annual research budget to be allocated to the faculties is fixed. This implies that the scores have a multiplicative degree of freedom. Multiplication of the scores by an arbitrary positive factor a means that the rewards per point are divided by a. What matters in the allocation is the ratio of any pair of scores. Hence, we asked the participants to compare every pair of output items and to estimate the ratio which seems to be fair for proportional funding, under each of the following objectives (distribution criteria):

1. The advancement of innovative pure and/or applied scientific research.
2. The advancement of innovative problem solving in the public or the private sector.
3. The documentation and dissemination of scientific knowledge and experience.
4. The coverage of the cost to produce the respective items.

We selected the following output items:

1. Refereed papers in outstanding international scientific journals.
2. Refereed papers in other scientific journals.
3. Books published by outstanding international scientific publishing companies.
4. Books published by other scientific publishers.
5. Contributions to refereed conference proceedings.
6. Doctoral dissertations, which usually take 4 or 5 years of innovative research.
7. Editorship of a scientific book or the proceedings of a conference.
8. Completion of a large third-stream project (more than US$ 250,000).
9. Completion of a medium-size third-stream project (US$ 25,000 to US$ 250,000).

University research projects may be funded by our National Science Foundation (the so-called second stream of financial resources) and by third parties (the third stream, mainly coming from industry). Although the cost of third-stream projects should in principle be covered by third parties, the acquisition of the project usually takes a considerable amount of first-stream effort. Note that the above list of output items is not exhaustive. The editorship of a scientific journal is also rewarded, for instance, but for such a regular task the editors may come to a special agreement with their faculties.

Let us illustrate the approach via the pairwise comparison of the items 1 and 3, the refereed papers and the books, both published via outstanding channels. Under the first objective (distribution criterion), counting the number of auto-references in several books in our own field and in adjacent fields, we might choose a ratio of 1:10 because the innovative contribution of a book seems to be more or less equivalent to the amount of innovation in 10 publications. Usually, however, these publications appeared in previous years so that the innovative contribution of books is predominantly the unification of the material scattered in the literature. On the other hand, many papers are never referred to, even if they appeared in outstanding scientific journals. Let us therefore set the ratio of papers and books to 1:3. Under the second objective (distribution criterion), the assessment is also non-trivial. The material in papers is usually presented in a concise manner so that it is not easily accessible for possible users. Books may be more mature so that a ratio of 1:3 could be appropriate. For the documentation and the dissemination of scientific knowledge, papers may be less effective and less mature than books so that a ratio of 1:5 could be plausible (we ignore that papers may be more up-to-date than books). Timewise, a paper (taking several months) is less expensive than a book (which may take several years). A ratio of 1:10 may be acceptable. Finally, we aggregate the above ratios via an aggregation procedure. The global ratio of a paper and a book, both published via outstanding channels, can be modelled as a weighted geometric mean of the above ratios. The weights reflect the relative importance of the objectives (distribution criteria) of output-based research funding. When we assign equal weights to the objectives the global ratio is given by

$$\sqrt[4]{\frac{1}{3}\times\frac{1}{3}\times\frac{1}{5}\times\frac{1}{10}}\approx\frac{1}{4.6}.$$

7.8 ASSIGNMENT OF SCORES TO RESEARCH OUTPUT ITEMS

We proposed four somewhat different elicitation procedures to find a proper set of scores for the respective output items under each of the relevant criteria. The exercise could interactively be carried out in the Group Decision Room (with GroupSystems, Ventana Corp., Phoenix, Arizona, see Dennis et. al., 1988) of the Faculty of Systems Engineering and Policy Analysis. The procedures, varying in the amount and the type of information to be supplied by the users, were derived from the Multiplicative AHP.

Pairwise comparisons with ratio information. This is exactly the procedure of the previous section, but now in a more formalized setting. Under each distribution criterion the participants are in principle invited to fill a 9×9 reciprocal matrix so that they have to supply the entries of the upper or the lower triangle only. The scores are given by the geometric row means of the respective matrices. The global scores assigned by the group of participants can be obtained via an aggregation procedure which is also based upon the calculation of geometric means.

Rating with ordinal information. If the users feel that it takes too much time and energy to carry out the pairwise comparisons they can merely try to rank the output items. Suppose, for instance, that the advancement of innovative pure and/or scientific research is supported in the following rank order: books by outstanding publishers > books by other publishers > dissertations > papers in outstanding journals > papers in other journals > editorship of a scientific book or conference proceedings > large third-stream projects > medium-size third-stream projects > contributions to refereed conference proceedings. The only additional piece of information we need is a rough estimate of the range ratio, that is, the ratio of the least important to the most important output item. Let it here be given by 1: 25. In the absence of any further information we assume that the ratios of the successive output items are equal. Hence, the scores assigned to the nine output items are on a geometric scale between 1 and 25, with the progression factor $\sqrt[8]{25} \approx 1.4953$. The resulting scores (the calculated ones between 1 and 25, as well as the normalized scores) are shown in Table 7.16. Even this simplified procedure, which is merely based upon ordinal information, enables the participants to discuss the relative importance of the criteria and the performance of the alternative output items under the respective criteria.

Rating with difference information. The users are invited to rank the output items and to put them on an arithmetic scale between the least and the most important output item. The other items must be interpolated on the scale in such a way that differences reflect importance ratios. The underlying assumption is that the users feel more comfortable with an arithmetic scale because it enables them to place an item exactly in the arithmetic middle of two other items if it is felt to be geometrically (proportionally) in the middle between them. We need again an estimate of the range ratio. Moreover the user has to choose the number of echelons on the scale. If the range ratio is 1:25 and if there are 13 echelons, for instance, the arithmetic scale represents in fact a geometric scale with the progression factor $\sqrt[12]{25} \approx 1.3077$. Table 7.17 shows the arithmetic scale 0, 1, 2, ..., 12, the corresponding geometric scale, and the normalized scores resulting from the procedure. Note that certain output items may coincide on the two scales. A comparison with Table 7.16 shows that the simple ranking procedure with ordinal information only produces practically the same scores.

Distribution of credit points. Here each participant is invited to distribute a given number of credit points over the alternative output items according to their relative importance. Thus, the credit points should appear in the ratios which are felt to be reasonable under the given distribution criterion.

Table 7.16 *Illustrative example. Scores assigned to the output items under the first-named distribution criterion, the advancement of innovative pure and/or applied scientific research. The assigment is based upon ordinal information. Note that normalization is a cosmetic operation only because the scores have a multiplicative degree of freedom.*

Output items under advancement of innov. research	Calculated scores	Normalized scores
Books published by outstanding int. sc. publishers	25.0000	0.340
Books published by other scientific publishers	16.7185	0.228
Doctoral dissertations	11.1803	0.152
Papers in outstanding scientific journals	7.4767	0.102
Papers in other scientific journals	5.0000	0.068
Editorship of scientific book or conf. proceedings	3.3437	0.046
Large third-stream projects	2.2361	0.030
Medium-size third-stream projects	1.4953	0.020
Contributions to refereed conference proceedings	1.0000	0.014

Table 7.17 *Illustrative example. Scores assigned to the output items under the first-named criterion, the advancement of innovative pure and/or applied scientific research. The items are placed on the arithmetic scale* 0, 1, 2, ..., 12, *which in fact represents a geometric scale between* 1 *and* 25 *with* 13 *echelons and henceforth the progression factor* 1.3077. *The scores are normalized in order to be more informative.*

Output items under advancement of inn. research	arithmetic scale	geometric scale	norm. scores
Books published by outstanding sc. publishers	12	25.0000	0.370
	11	19.1181	
Books published by other scientific publishers	10	14.6200	0.216
Doctoral dissertations	9	11.1803	0.165
	8	8.5499	
Papers in outstanding scientific journals	7	6.5383	0.100
	6	5.0000	
Papers in other journals & Editorship of book	5	3.8236	0.056
	4	2.9240	
Large & Medium-size third-stream projects	3	2.2361	0.033
	2	1.7100	
	1	1.3077	
Contributions to refereed conference proceedings	0	1.0000	0.015

In the GDR session of November 17, 1997, twelve members of the project team jointly evaluated the nine output items under one objective or distribution criterion only, the advancement of pure and/or scientific research. There was not enough time to do more. First, they ranked the items in a decreasing order of importance and they supplied the range ratios. This procedure was felt to be simple and adequate. Thereafter, they distributed 100 credit points over the output items. Because a normalization routine was not available this procedure was felt to be inadequate, at least under the given pressure of time (a user-friendly routine would allow the members to distribute roughly 100 credit points, and it would automatically normalize the assigned scores thereafter without affecting their ratios). Finally, they applied pairwise comparisons with ratio information. Under the pressure of time this procedure was felt to be confusing because it could easily lead to inconsistencies and cyclic judgement. Nevertheless, the three procedures gave practically the same results. This is shown in Table 7.18, and it is also clear from the Spearman rank correlation factors: 0.85 for pairwise comparisons versus rating with ordinal information, 0.96 for pairwise comparisons versus distribution of credit points, and 0.81 for rating with ordinal information versus distribution of credit points.

Table 7.18 *Preliminary results of the experiment in the Group Decision Room. Normalized scores assigned to the output items under the first-named criterion, the advancement of pure and/or applied scientific research, according to three different procedures.*

Output items under the criterion of advancement of innovative research	Pairwise comparisons	Rating with ordinal info	Distribution of credit points
Papers in outstanding scientific journals	0.33	0.19	0.28
Books published by outstanding int. sc. publishers	0.18	0.16	0.16
Doctoral dissertations	0.12	0.13	0.11
Large third-stream projects	0.10	0.09	0.14
Books published by other scientific publishers	0.09	0.11	0.06
Papers in other scientific journals	0.07	0.13	0.11
Editorship of scientific book or conf.proceedings	0.04	0.05	0.05
Medium-size third-stream projects	0.04	0.08	0.03
Contributions to refereed conference proceedings	0.03	0.05	0.06

As we remarked in the previous section, the scores in Table 7.18 have a multiplicative degree of freedom so that only the ratio of any pair of scores is meaningful. Papers in outstanding journals and books from outstanding publishers, for instance, should roughly be rewarded in the ratio 2 : 1, at least in the opinion of the participants.

Although the project has not yet been completed, it brought up many issues, both during and after the session in the Group Decision Room. We had many discussions, not only with project-team members but also with colleagues and university administrators, and particularly with the Allocation Committee which designed the system and now supervises its performance. Obviously, there are several different and contradictory views on the significance of the output items. We briefly summarize the comments here.

- Output-based research funding is not necessarily innovative because the transition from the old input-based system must be smooth. Output-based allocation systems tend to be conservative because the positions of the tenured scientific staff must be maintained (see also Fandel and Gal, 1997). Of course, one could design an ideal system, the endpoint of a controlled transition process which may take several years, but it remains questionable whether the endpoint will ever be reached.

- In the 1995 – 1998 versions of the Delft allocation system, some faculties (Electrical Engineering, for instance) received more rewards per point than others (Mathematics and Computer Science, for instance). The argument was that they needed more technical personnel and equipment for the production of their research output. They may have played a power game, however. This is true for many allocation systems: the scores are not always the result of a clean and transparent procedure.

- For university administrators, many research output items are financially not interesting. Books, doctoral dissertations, and editorships are relatively rare so that they hardly affect the allocation of the research budget. In the Mathematics Department, for instance, the bulk of the output is given by the articles in outstanding and other journals, in the Computer Science Department by the contributions to conference proceedings and by the so-called business publications. The last-named items are research-related publications in general newspapers and journals (in English or in the national language), extended project proposals, consultancy reports, manuals for computer programmes, etc. This category has not been considered in the above experiments.

- Scientists are not encouraged to write books, see also the ratios of the scores assigned to books and papers in Table 7.18. This may be a structural phenomenon. The Science Citation Index can be used to trace the impact of scientific papers in refereed journals (the most important output items!), but it does not properly show the impact of books. The same is true for contributions to conference proceedings. They have a status which is much lower than the status of papers in journals.

- It is not clear what the decision makers have in mind when they compare the output items. What do they have in mind when they compare papers and books, for instance? A small number of pioneering papers, frequently cited in the journals, or the mass of rarely cited papers with minor innovations? Some popular textbooks, a few thorough overviews of the current state-of-the-art, or some special monographs?

- Doctoral dissertations should not be rewarded at all. If they are equivalent to two or three publications in outstanding journals (unfortunately, this is not always the case) the system could reward the publications as soon as they appear in the scientific literature.

- The same argument holds for the projects. If they contribute to the advancement of pure and/or applied research (and university projects should do so) one only has to postpone the remuneration until the spin-off appears in the scientific literature. No wishful thinking about the significance of the projects, only firm proofs should be accepted!

- Services to the scientific community are seriously underestimated. Editors of books, journals, and conference proceedings are (meagerly?) remunerated in the allocation system. The members of the editorial boards are not. No compensation is offered for the humble tasks of the anonymous referees who have to carry the load of the publication activities. The load will certainly increase by output-based research funding.

- The total annual research budget is fixed so that the allocation system will presumably enhance the competitive behaviour of the faculties. This may lead to undesirable strategies such as the publication of two ideas in two separate papers, not in one paper jointly, even if it would have been beneficial for the readers.

On the short term, output-based research funding does not seem to affect the relative position of the faculties because tenured scientific staff cannot easily be moved. On the long term, however, it may be effective because the annual publication of the financial results will permanently put the scientific staff under pressure. Staff members may accordingly increase their output, but they may also decide to leave the faculty, possibly via early retirement.

Scientists with a project in full swing will benefit from output-based research funding. Their instruments work properly, their computer programs have been debugged, and their choice of a research direction appears to be fortunate. They may produce a significant amount of high-level output. The system may be disastrous for scientists who work on the boundary of the established disciplines, who explore totally new research directions, and who challenge the widely accepted paradigms of their field, unless they are protected by their faculties. The Allocation Committee does not have the task, however, to verify whether the faculties have a proper mechanism (high-quality science committees, for instance) to protect such entrepreneurial research.

The allocation system in Delft has recently been revised. The rewards per point will be the same for all faculties, and doctoral dissertations will be rewarded with a score of 12 points instead of 10 because the salaries of the research assistants have been increased. The other scores have been frozen. Frequently adjusted under the pressure of the faculties, they will remain unchanged during two or three years in order to stabilize the system. There are several additional rules (lump sums to the faculties) whereby the transition to output-based research funding has been lubricated. These additional rules will gradually be dropped so that the system can eventually show its full effect. In the current transition period the Allocation Committee did not express a keen interest in a new evaluation of the scores.

In the course of 1998, two research assistants carried out a full-scale experiment in our Department of Statistics, Stochastics, and Operations Research. They submitted a questionnaire to the scientific staff members, asking them to evaluate the current research output items under the four criteria mentioned in Section 7.7, according to the method of rating with ordinal information. There were ten respondents: two senior professors, four associate professors, and four research assistants. The results are summarized in Table 7.19. Note that the output items are the relevant ones for the Department SSOR in the current allocation system, whereas the items in the previous experiment of the DAS Working Group had been chosen rather freely because the participants came from a variety of universities.

Table 7.19. *Full-scale experiment. Normalized scores of nine research output items under four criteria by ten members of the Department SSOR of the Faculty of Information Technology and Systems.*

Criteria for output-based funding: advancement of	pure/applied research	problem solving	knowledge dissemination	cost coverage	normalized scores
Criterion weights	0.50	0.18	0.17	0.15	
Papers in outstanding journals	0.31	0.25	0.28	0.21	0.27
Books by outstanding publisher	0.22	0.14	0.26	0.23	0.22
Doctoral dissertations	0.18	0.13	0.08	0.14	0.16
Contributions to book	0.10	0.08	0.12	0.13	0.11
Short contributions to book	0.05	0.05	0.07	0.05	0.05
Contr. to refereed proceedings	0.07	0.11	0.11	0.07	0.08
Reports to third parties	0.03	0.15	0.02	0.10	0.05
Business publications	0.01	0.03	0.04	0.06	0.05
Abstracts	0.03	0.06	0.03	0.02	0.02

There is a remarkable similarity between the scores assigned to the common items in the two experiments, see the Tables 7.18 and 7.19. Using the multiplicative degree of freedom of the results in a rough and easy manner, one could assign a score of 5 to papers in outstanding journals, 4 to books by outstanding publishers, 3 to doctoral dissertations, and 2 to contributions to conference proceedings. This sharply contradicts the current assignment of scores (5, 10, 10 or 12, and 3 respectively, see Section 7.7).

The Allocation Committee, confronted with the outcome of the experiments, pointed at two shortcomings. First, the participants came from a particular field of research, whereas the system had been set up for a wide range of engineering faculties with totally different research experiences. Second, young researchers tend to overestimate the impact of papers, and they underestimate the innovative unification as well as the maturity of books. It is possible that a generation conflict emerges here. The members of the Allocation Committee have a long and outstanding career in scientific research and in science management. Books and dissertations are also highly valued by the SSOR members who published one or more books. It is also possible that the respective parties start from totally different viewpoints. The Allocation Committee has the objective to reward the faculties in their totality, whereas the staff members have the impression that the assignment of scores is an assessment of their individual research activities. For the time being, the experiments brought up several interesting points of discussion.

REFERENCES TO CHAPTER 7

1. Aristotle, "*Ethics*", in the translation by J. Warrington. Dent and Sons, London, 1963.

2. Beckerman, W., "*An Introduction to National Income Analysis*". Weidenfeld and Nicolson, London, 1980.

3. Dennis, A., George, R.J.F., Jessup, L.M., Nunamaker, J.F., and Vogel, D.R., "Information Technology to support Electronic Meetings". *MIS Quarterly* 12, 591 - 624, 1988.

4. Deutsch, M., "Equity, Equality, and Need: What Determines which Value will be Used as the Basis of Distributive Justice?" *Journal of Social Issues* 31, 137 - 149, 1975.

5. Deutsch, M., "*Distributive Justice*". Yale University Press, New Haven, 1985.

6. "*Eurostat Yearbook* 1995, *a Statistical Eye on Europe* 1983 - 1993". Office for the Publications of the European Communities, Luxemburg, 1995.

7. Fandel, G., und Gal, T., "Umverteilung der Mittel für Lehre und Forschung unter den Universitäten". Diskussionsbeitrag 248, FB Wirtschaftswissenschaft, Fernuniversität Hagen, D-58084 Hagen, Deutschland, 1997.

8. Fishburn, P.C., and Sarin, R.K., "Fairness and Social Risk I: Unaggregated Analysis". *Management Science* 40, 1174 - 1188, 1994.

9. Kasperson, R.E. (ed.), "*Equity Issues in Radioactive Waste Management*". Oelschlager, Gunn, and Hain, Cambridge, 1983.

10. Linnerooth-Bayer, J., Davy, B., Faast, A., and Fitzgerald, K., "Hazardous Waste Cleanup and Facility Siting in Central Europe: the Austrian Case". Technical Report GZ 308.903/3-43/92, IIASA, Laxenburg, Austria, 1994.

11. Lootsma, F.A., "Numerical Scaling of Human Judgement in Pairwise-Comparison Methods for Multi-Criteria Decision Analysis". In G. Mitra (ed.), "*Mathematical Models for Decision Support*". Springer, Berlin, 1988, pp. 57 – 88.

12. Lootsma, F.A., "The French and the American School in Multi-Criteria Decision Analysis". *RAIRO/Recherche Opérationnelle* 24, 263 - 285, 1990. A short version appeared in A. Goicoechea, L. Duckstein, and S. Zionts (eds.), "*Multiple Criteria Decision Analysis*". Springer, New York, 1992, pp. 253 – 268.

13. Lootsma, F.A., Mensch, T.C.A., and Vos, F., "Multi-Criteria Analysis and Budget Reallocation in Long-Term Research Planning". *European Journal of Operational Research* 47, 293 – 305, 1990.

14. Lootsma, F.A., and Bots, P.W.G., "The Assignment of Scores for Output-Based Research Funding". To appear in the *Journal of Multi-Criteria Decision Analysis*, 1999.

15. Lootsma, F.A., Ramanathan, R., and Schuijt, H., "Fairness and Equity via Concepts of Multi-Criteria Decision Analysis". In T.J. Stewart and R. van den Honert (eds.), *"Trends in Multi-Criteria Decision Making"*. Springer, Berlin, 1998, pp. 215 – 226.

16. Raan, A.F.J. van, "Advanced Bibliometric Methods to Assess Research Performance and Scientific Development". *Research Evaluation* 3, 151 – 166, 1993.

17. Raan, A.F.J. van, "Advanced Bibliometric Methods as Quantitative Core of Peer Review Based Evaluation and Foresight Exercises". *Scientometrics* 36, 397 – 420, 1996.

18. Messick, D.M., and Cook, K.S. (eds.), *"Equity Theory, Psychological and Sociological Perspectives"*. Praeger, New York, 1983.

19. Young, H.P., *"Equity in Theory and Practice"*. Princeton University Press, New Jersey, 1994.

CHAPTER 8

SCENARIO ANALYSIS

Strategies and scenarios are usually designed by strategic planners who prepare a decision which is irreversible for a long period of time. First, the factors are identified which affect the economic, political, and social developments until a predetermined horizon and which are beyond the decision maker's control. Because these factors are mostly interconnected, the hypothetical future developments can be bundled into a small number of coherent scenarios. Usually, the planners design a trend-following scenario where the current developments are continued without interruptions (business as usual), an optimistic scenario where the developments proceed in a more upward direction, and a pessimistic scenario with a less upward or even downward direction of the developments. Each strategy has certain desirable and/or undesirable consequences within the context of the respective scenarios. In choosing a strategy, the decision makers have to evaluate and to weigh these consequences and to take into account the scenario probabilities. For the benefits and the pitfalls of scenario analysis, see Leemhuis (1985) and de Geus (1988), the pioneers at Royal Dutch/Shell in the nineteen seventies, and Mintzberg (1994).

8.1 STRATEGIES AND SCENARIOS

The evaluation of the strategies within the scenarios under consideration is a complicated process, usually prepared in a network of policy committees and advisory councils. Mostly, these groups are expected to suggest a preferred strategy; the recommendation may be based upon a majority decision or on group consensus. The capacity of human beings to judge a large number of consequences in a variety of hypothetical circumstances is so limited, however, that formalized methods may be necessary for the group to arrive at a conclusion. These methods are indeed available. Within the framework of a particular scenario, MCDA can be used to evaluate the strategies under a number of viewpoints or criteria. Combination of the results by means of plausible scenario probabilities could possibly lead to the final choice of a particular strategy. The present chapter describes our experiences since the late eighties with such a formalized analysis in the national electricity sector of The Netherlands.

The dominant problems are well-known. On the demand side, there had been an annual increase varying between 0% and 7% since 1945. Difficulties arose on the supply side. In order to cover the base-load electricity demand, one had the choice between four primary-energy carriers: coal, oil, natural gas, and uranium, each with far-reaching and irrevocable consequences for the national economy, the environment, the strategic independence in energy matters, and the national safety. The energy debate in The Netherlands was and is further conditioned by an abundant supply of domestic natural gas, a huge refinery capacity in the harbour of Rotterdam, strong public opposition against nuclear energy, and an unexpected environmental damage due to fossil fuels. Moreover, the Low Countries will be in the front line if the greenhouse effect leads to a sea-level rising. The problem of how to choose a fuel strategy is not typical for The Netherlands, however. In many countries, long-term plans for the national electricity supply are frustrated by similar opposite threats.

In the nineteen eighties, our country was for more than 80% dependent on oil and natural gas. In the electricity sector, some 50% of the demand was covered by natural gas, 40% by coal, and 7% by domestic nuclear energy. The dependence on fossil fuels was increasingly felt to be undesirable, but a further diversification towards nuclear energy was not easily acceptable either. Solar energy (too expensive) and wind energy (too whimsical) were not sufficiently promising, at least for several decades ahead. After the Chernobyl disaster of 1986, however, the national government avoided the authorization of investments in nuclear power plants (the two small nuclear plants in Dodewaard and Borssele dated from the seventies), so that the national energy policy was practically in a stalemate. Figure 8.1 shows the dislocation of the power plants in The Netherlands in the early nineties. There were four electricity companies working under the umbrella of the Cooperating Electricity Producers (SEP) in Arnhem. The national production capacity, roughly 14.400 MW, did not completely cover the annual consumption, roughly 65 $\times 10^{12}$ kWh. Almost 15% of the demand was covered by imports from the neighbouring countries: Belgium, France, and Germany. This was due to the decisions in the late eighties. The electricity companies, in their attempts to keep the nuclear option open as long as possible, proposed to construct gas-fuelled power plants only and to cover the peak demand by imports under long-term contracts with foreign companies. In doing so, they enabled the national government to avoid the authorization of delicate investment decisions. The nuclear issue had been moved beyond the planning horizon of ten years. In the present chapter we will extensively describe the political stalemate as well as the way out, via an alternative strategy which was originally not available in the scenario studies of the Ministry of Economic Affairs.

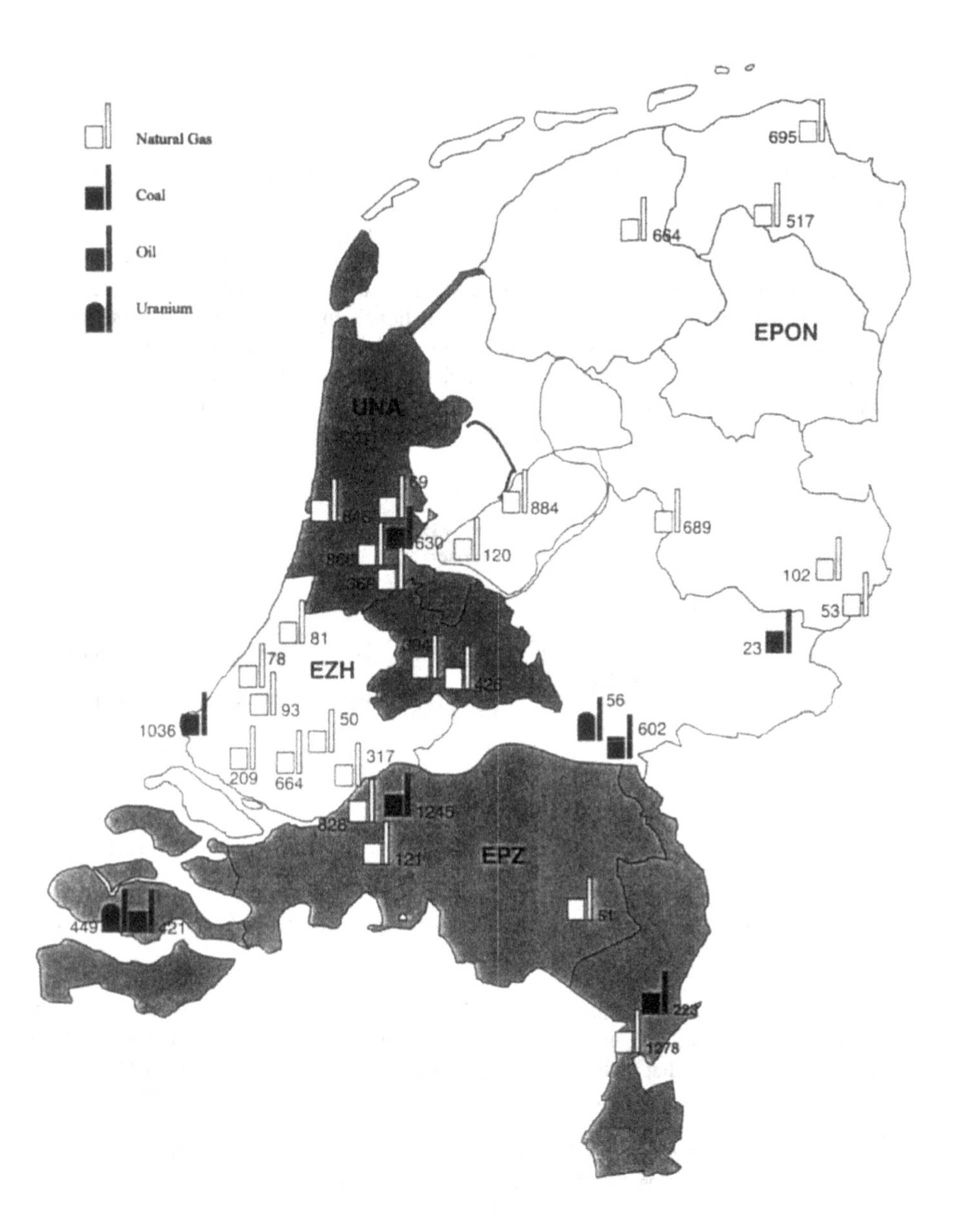

Figure 8.1 *Dislocation of the power-generating capacity of The Netherlands in the early nineties. There were four electricity-producing companies (EPON, EPZ, EZH, and UNA). The numbers on the map represent the capacities of the respective plants in MW. Source: The Energy Report* 1994.

8.2 THE NATIONAL ENERGY MODEL

In the late seventies the Ministry of Economic Affairs commissioned the Energy Study Centre (Petten, North-Holland) to design an energy model of the national economy, in order to support the development of a long-term strategy for the national energy supply. It is beyond the scope of this volume to present a detailed description of the medium-size linear-programming model. We restrict ourselves to a brief explanation of the structure exhibited in Figure 8.2. The design of the national energy model in Petten closely followed the ideas developed at the Brookhaven National Laboratory (Long Island, USA) in the early seventies, when politicians and the general public became aware of the scarcity of primary energy carriers. It is mainly a flow model where energy carriers move from the suppliers via the conversion units to the end users.

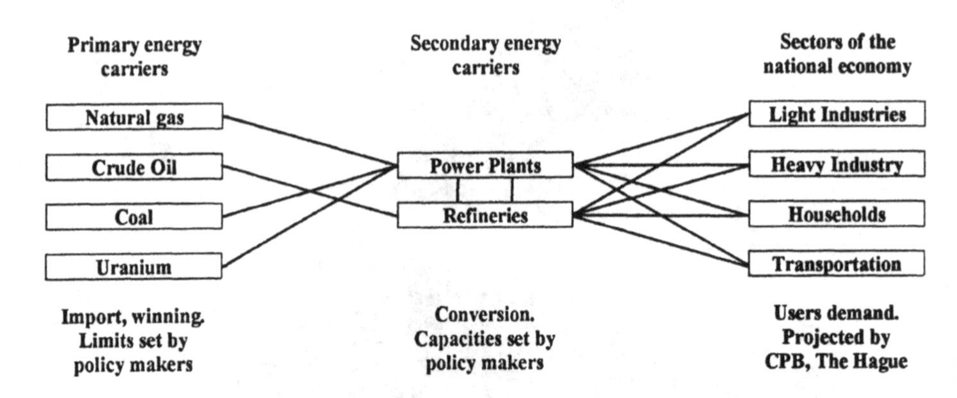

Figure 8.2. *Overview of the national energy model in Petten (North-Holland).*

At the input side one finds the import and the winning of the primary energy carriers: natural gas, crude oil, coal, and uranium. Note that the model has been designed at a high aggregation level so that "winning of natural gas" represents winning in a large region, not an individual source of natural gas in Groningen or in the North Sea. Next, the primary energy carriers flow to the units where they are converted into secondary energy carriers such as electricity or various types of petrol. These products, which are obviously ready for immediate consumption, flow to various sectors of the national economy: light and heavy industry, households, transportation, and the like.

The interesting feature of energy models is that they incorporate the national energy policy via the constraints on the flows. At the input side there are limits on the import and the winning of primary energy carriers. Our country maintains its energetic independence by limits on the import of crude oil and on the depletion of domestic natural gas. In the conversion domain there are controversial limits on the capacity of nuclear power plants. At the demand side, however, one has to work with certain assumptions and with the resulting projections of the future energy demand. The energy model in Petten is fed with the demand projections given by the Central Planning Bureau in The Hague. In fact, Petten has a static network flow model where the data correspond to the hypothetical situation in a future sample year such as 2000 or 2010. The flows are not uniquely determined by the capacity limitations and the demand requirements. The degrees of freedom are therefore used to find a solution which minimizes the total costs for maintenance, fuels, operations, and capital in the sample year.

In principle, there are several non-linearities in the model, particularly in the total cost function. Moreover, the model is complicated by the relationships between the incoming and the outgoing flows at the conversion nodes. Nevertheless, the model has been reduced to a general linear-programming problem. We can now briefly summarize what it actually does. For a pre-specified year, using as input

- projections of the energy demand in various sectors of the economy,
- projections of prices of primary energy carriers,
- estimates of investment costs of (new) energy technologies,
- projections of supply restrictions, capacity limitations, and conversion efficiencies,

the linear-programming model (in the eighties roughly 400 constraints and 500 variables) calculates an optimal mix of secondary energy carriers and technologies. The projections depend heavily on the scenario under consideration.

8.3 THE NATIONAL ENERGY OUTLOOK 1987

We became deeply involved in scenario analysis when the National Energy Outlook 1987 appeared, a study carried out by the Energy Study Centre in Petten (North-Holland) to the order of the Ministry of Economic Affairs. Within the context of three scenarios characterized by low, medium, and high economic growth, the Centre evaluated three strategies to cover the base-load electricity demand (not completely, but to a high degree): a nuclear, a coal, and a natural-gas strategy. The consequences of the nine strategy-scenario combinations, calculated via the national energy model which generated cost-optimal solutions in the years 2000 and 2010 under scenario-dependent capacity limitations and demand requirements, were not only expressed in monetary units (total cost, price per kWh in the respective sample years). The report also showed the resulting air pollution (SO_2, NO_x, CO_2 emissions in 2000 and 2010), the remaining gas reserves, the dependence on oil and natural gas, and the amount of nuclear power, in order to characterize the future energy position of The Netherlands following from the respective energy-supply strategies.

It was by no means clear whether a decision maker could reasonably arrive at a well-balanced conclusion on the basis of the flood of data in the National Energy Outlook 1987. In order to prepare the ground for a formalized decision analysis, we first summarized the information as much as possible. Table 8.1 shows the three scenarios. Each of them was based upon a coherent set of hypothetical growth rates assigned to the dominant factors behind the scenarios: the crude-oil price (which affects all other energy prices), the Gross National Product, and the electricity demand. Moreover, there were three strategies which differed in the relative amount of nuclear, coal, or natural-gas capacity installed to satisfy the base-load electricity demand. Note that in each strategy the peak-load units were fuelled by natural gas so that demand fluctuations during a 24 hours period could rapidly be followed. Table 8.2 shows the performance of the strategies within the context of each of the scenarios, in the sample years 2000 and 2010. There were nine quantitative criteria, with the following monetary or physical units to express the performance of the alternative strategies.

1. Production costs in cents/kWh, to measure the competitive position of the national industry with respect to the industries abroad (1 cent = Dfl 0.01 ≈ US$ 0.05).

2. The total cost of the national electricity supply, as a percentage of the Gross National Product.

3. SO_2 emissions in thousands of tons above the limit of 140,000 tons/year which was laid down in international agreements.

4. NO_x emissions in thousands of tons above the limit of 140,000 tons/year which was laid down in international agreements.

5. Domestic natural-gas reserves in years of consumption after 2000 and 2010 respectively, in order to measure the energy dependence of The Netherlands.

6. Profits on domestic natural gas in billions of Dutch guilders (Dfl 1.00 ≈ US$ 0.50).

7. Percentage of oil and natural gas in the national energy consumption, in order to measure the diversification in the energy system.

8. Installed nuclear capacity in MW, in order to measure the unsafety of the national energy system.

9. CO_2 emissions in GtC/year, in order to measure the Dutch share in the causes of a climatic change (the greenhouse effect, possibly followed by a sea-level rising).

Under the criteria 1 – 4 and 7 – 9 the indicators (costs, emissions, nuclear capacity) were to be minimized. Under the criteria 5 and 6, on the other hand, the indicators (delayed consumption, profits) were to be maximized.

The four authors of the paper by Lootsma et al. (1990) carried out the experiments to be described in the next section. Two of them were from the Energy Study Centre and two from the Delft University of Technology. They decided, first, to imagine themselves in the role of decision makers in order to test the MCDA methodology and the supporting software. Their initial analysis consisted of three parts: MCDA within the context of the low, the medium, and the high scenario respectively. Within each scenario they were confronted with three decision alternatives to be evaluated under nine performance criteria.

Table 8.1 *Strategy-scenario combinations in the National Energy Outlook* 1987 *and the corresponding diversification of the electricity production in the sample years* 2000 *and* 2010. *The diversification in the energy-supply strategies (enhanced use of nuclear energy, coal, or natural gas) is due to the choice of fuels (nuclear energy, coal, natural gas, other fuels) to cover the base-load electricity demand. Peak-load units are fuelled by natural gas in all strategies. The entries in the columns 5 – 7 represent the contributions of the respective fuels in percentages.*

2000

Scenario	Factors	%	Type of fuel	Nuclear str.	Coal str.	Nat.-gas str.
	Annual increase of		Nuclear	21	6	6
Low	Crude-oil price	1.6	Coal	48	66	53
	GNP	1.5	Natural gas	27	23	36
	Electricity demand	0.9	Other fuels	4	4	4
	Annual increase of		Nuclear	26	5	5
Medium	Crude-oil price	2.8	Coal	48	73	43
	GNP	2.5	Natural gas	22	19	49
	Electricity demand	1.9	Other fuels	4	3	3
	Annual increase of		Nuclear	32	4	4
High	Crude-oil price	3.7	Coal	47	78	38
	GNP	3.5	Natural gas	18	15	55
	Electricity demand	2.7	Other fuels	3	3	3

2010

Scenario	Factors	%	Type of fuel	Nuclear str.	Coal str.	Nat.-gas str.
	Annual increase of		Nuclear	54	0	0
Low	Crude-oil price	1.6	Coal	22	69	49
	GNP	1.5	Natural gas	17	23	43
	Electricity demand	0.9	Other fuels	7	8	8
	Annual increase of		Nuclear	56	0	0
Medium	Crude-oil price	2.8	Coal	23	79	39
	GNP	2.5	Natural gas	15	14	54
	Electricity demand	1.9	Other fuels	6	7	7
	Annual increase of		Nuclear	31	0	0
High	Crude-oil price	3.7	Coal	21	81	33
	GNP	3.5	Natural gas	14	13	61
	Electricity demand	2.7	Other fuels	4	6	6

Table 8.2 *Consequences of the strategy-scenario combinations analyzed in the National Energy Outlook* 1987. *Performance tableau of three energy-supply strategies (nuclear, coal, natural gas) under nine performance criteria in the sample years* 2000 *and* 2010, *within the framework of three economic-growth scenarios (low, medium, high).*

Scenario	Criteria	Nuclear strategy		Coal strategy		Natural-gas strategy	
		2000	2010	2000	2010	2000	2010
Low	Comp. position	9.4	10.3	9.4	10.9	9.6	11.2
	Total cost	21.8	19.6	21.8	19.7	21.8	19.8
	SO_2 emissions	40	50	50	80	40	60
	NO_x emissions	240	220	250	270	250	270
	Gas reserve	27	17	26	16	24	14
	Profits on gas	22.3	15.7	22.3	16.7	23.0	17.8
	Diversification	77	70	78	75	81	80
	Safety	1830	6550	530	50	530	50
	CO_2 emissions	0.04	0.04	0.04	0.04	0.04	0.05
Medium	Comp. position	10.1	10.8	10.3	11.5	11.1	13.1
	Total cost	20.8	19.0	20.8	19.2	20.9	19.3
	SO_2 emissions	80	80	100	140	60	80
	NO_x emissions	300	280	320	350	310	340
	Gas reserve	27	17	26	16	21	11
	Profits on gas	28.3	21.3	28.3	21.8	31.1	26.6
	Diversification	71	65	72	67	80	79
	Safety	3130	9150	530	50	530	50
	CO_2 emissions	0.05	0.05	0.05	0.06	0.05	0.05
High	Comp. position	10.9	11.5	11.4	13.3	12.6	16.1
	Total cost	22.2	22.2	22.3	22.5	22.4	22.8
	SO_2 emissions	90	120	120	200	60	110
	NO_x emissions	320	340	350	430	350	420
	Gas reserve	25	15	25	15	18	8
	Profits on gas	33.8	28.6	34.1	29.8	38.7	38.7
	Diversification	70	63	71	67	83	81
	Safety	4430	13050	530	50	530	50
	CO_2 emissions	0.05	0.05	0.05	0.07	0.05	0.06

8.4 FIRST EXPERIMENT: TESTING THE METHODOLOGY

The experiment was not easy to carry out, and it did not follow the guidelines of the previous chapters. Although there were many data to characterize the performance of the alternative strategies, the authors did not attempt to frame the problem. They used the Multiplicative AHP in the form of the late eighties (with the arithmetic-mean aggregation rule and with simultaneous scale variations for criteria and alternatives, see Chapter 4), and they expressed their comparative judgment directly in verbal terms, after careful inspection of the data in Table 8.2.

Table 8.3 shows the terminal scores of the strategies in the three scenarios, on the two scales normally used in the eighties. First, there are the terminal scores which follow from the judgemental statements of the individual authors. Next, one finds the group terminal scores, the result of attempts to arrive at a compromise solution throughout the analysis, when the group acts as a single decision maker (see Chapter 6).

Table 8.3 *Terminal scores of the energy-supply strategies within the economic-growth scenarios. The scores are due to verbal judgemental statements converted into numerical values on two different geometric scales.*

Scenario	Author	Energy-supply strategies to satisfy the base-load electricity demand					
		Scale parameter $\gamma = 0.72$			Scale parameter $\gamma = 1.44$		
		Nuclear	Coal	Nat. gas	Nuclear	Coal	Nat. gas
Low	1	0.45	0.23	0.33	0.49	0.12	0.39
	2	0.49	0.26	0.25	0.72	0.13	0.16
	3	0.06	0.42	0.52	0.00	0.50	0.50
	4	0.32	0.36	0.31	0.16	0.47	0.36
	group	0.29	0.34	0.37	0.07	0.43	0.50
Medium	1	0.41	0.22	0.37	0.37	0.12	0.52
	2	0.54	0.20	0.26	0.77	0.07	0.16
	3	0.08	0.44	0.48	0.00	0.63	0.37
	4	0.37	0.36	0.27	0.26	0.45	0.29
	group	0.31	0.31	0.39	0.09	0.34	0.57
High	1	0.44	0.21	0.35	0.47	0.09	0.44
	2	0.53	0.20	0.28	0.75	0.06	0.19
	3	0.07	0.50	0.43	0.00	0.70	0.30
	4	0.35	0.35	0.30	0.18	0.47	0.35
	group	0.30	0.33	0.37	0.07	0.39	0.54

Author 1 preferred the nuclear strategy in almost all circumstances. Rank reversal between the nuclear and the natural-gas strategy occurred in the medium scenario only. Note that the Multiplicative AHP in its present form would not produce rank reversals under variations of the scale parameter (see Chapter 4). Author 2 preferred the nuclear strategy in all scenarios and on both scales. Author 3 eventually preferred the coal strategy over the natural-gas strategy, despite the rank reversal in the medium scenario. Author 4 eventually preferred the coal strategy over the nuclear strategy. The leading criteria of the group were safety (group compromise weight 0.35 and 0.67 on the respective scales) and minimization of the air-polluting emissions. Remarkably enough, the group preferred the natural-gas strategy in all scenarios and on both scales, although it was not consistently preferred by any of the authors individually. It was typically a compromise strategy. A coalition of the authors 1, 2, and 4, however, might have led to a majority decision in favour of the nuclear strategy. In general, much depends on the objectives of the group: do they want to present the undiluted viewpoint of a majority, or do they want to attain consensus with maximal support?

A striking phenomenon was that the choice of a strategy did not depend on the scenarios. There may have been two reasons for this. First, scenario analysis is (too?) difficult because one has to weigh the consequences of the strategies within the context of an imaginary situation, ten or twenty years ahead. Second, the scenarios were not sufficiently divergent, so that the corresponding images of the future had a significant amount of overlap. Merely based upon trend extrapolations, the scenarios did not sketch hypothetical worlds with different political structures, for instance.

Attempts to frame the decision problem brought up serious differences of opinion. Safety of the energy system was linked to the nuclear capacity in The Netherlands. For some authors the coal and the natural-gas strategy were accordingly much safer than the nuclear strategy. For others, however, the difference in safety was marginal, given the nuclear capacity in the neighbouring countries, installed along the Dutch border.

8.5 SECOND EXPERIMENT: FULL-SCALE ANALYSIS

In June 1988, we presented the results of the above experiment to the Working Group Integral Energy Scenarios. Established in 1980 by the Energy Study Centre, just before the Broad Public Discussion in The Netherlands on the threats and challenges of nuclear energy, the Group eventually consisted of 35 members who were professionally involved in studies on energy supply and environmental protection. They were employed by universities, research institutes, ministries, and by oil, gas, and electricity companies, but they contributed in person to the discussions on explorations of the future. After the presentation the Group decided to carry out a similar experiment, albeit in the medium scenario only. They agreed that the low and the high scenario were not sufficiently divergent to warrant the additional effort. Eventually, 20 members returned their pairwise comparisons via the special questionnaires designed for the experiment. The results, summarized in Table 8.4, were extensively discussed with the Working Group in the autumn of 1988.

Table 8.4 shows the terminal scores of the alternative strategies on two different scales. The rows 1 – 20 contain the terminal scores derived from the pairwise comparisons expressed by the decision makers individually. The last row exhibits the compromise solution when the group is supposed to act as a single decision maker, not as a collection of individuals. The members 8 and 18 only compared the alternatives under the respective criteria; hence, there was not enough information to calculate their terminal scores. Member 3 evaluated the criteria only; his judgement is also included in the compromise solution. Rank reversal due to scale variations only occurred in the results of member 1. In the Working Group, however, this phenomenon undermined the credibility of the proposed methodology. Nine members preferred the nuclear strategy, six the natural-gas strategy, and there was only one member in favour of the coal strategy. The group compromise solution exhibited the same order of preference, with the highest terminal score for the nuclear strategy. The results were easy to understand from inspection of the criterion weights (not shown in Table 8.4). High weights were generally assigned to the minimization of SO_2 and NO_x emissions and to safety (group compromise weights roughly 0.19 and 0.25 each on the respective scales).

Table 8.4 *Evaluation of the energy-supply strategies within the framework of the medium economic-growth scenario by members of the Working Group Integral Energy Scenarios.*

Member	Energy-supply strategies to satisfy the base-load electricity demand					
	Scale parameter $\gamma = 0.72$			Scale parameter $\gamma = 1.44$		
	Nuclear	Coal	Nat. gas	Nuclear	Coal	Nat. gas
1	0.41	0.31	0.28	0.18	0.41	0.41
2	0.44	0.23	0.33	0.67	0.11	0.23
3	-	-	-	-	-	-
8	-	-	-	-	-	-
10	0.50	0.17	0.33	0.73	0.02	0.24
11	0.09	0.41	0.51	0.00	0.47	0.53
12	0.57	0.13	0.30	0.88	0.02	0.10
13	0.38	0.29	0.34	0.42	0.34	0.24
15	0.45	0.12	0.43	0.59	0.01	0.39
17	0.26	0.25	0.49	0.09	0.14	0.78
18	-	-	-	-	-	-
19	0.36	0.26	0.38	0.04	0.17	0.78
20	0.35	0.22	0.43	0.05	0.13	0.82
21	0.18	0.56	0.26	0.02	0.83	0.15
23	0.68	0.12	0.20	0.95	0.01	0.04
25	0.47	0.25	0.28	0.71	0.12	0.17
28	0.52	0.31	0.17	0.79	0.17	0.04
30	0.26	0.21	0.54	0.04	0.06	0.91
31	0.60	0.18	0.22	0.90	0.04	0.06
32	0.17	0.25	0.58	0.02	0.12	0.86
group	0.43	0.26	0.31	0.52	0.16	0.32

Several members of the Working Group disliked the scale sensitivity of the weights and scores as well as the occurrence of rank reversals, but in their critical comments they concentrated on the choice of the criteria, which was more or less dictated by the National Energy Outlook 1987. The three emission criteria, for instance, worked in the same direction: they supported the nuclear strategy at the expense of the coal and the natural-gas strategy which are based upon fossil fuels. The set of nine criteria could have been reduced to four:

1. Cost minimization (total cost of electricity supply + profits on domestic natural gas).

2. Minimization of air pollution (SO_2, NO_x, and CO_2 emissions).

3. Maximization of independence (comp. position + gas reserves + diversification).

4. Maximization of safety (minimization of nuclear capacity).

Nevertheless, the Working Group did not expect a significant change of weights and scores by an analysis on the basis of the new set of criteria. Minimization of air pollution and maximization of safety were highly valued in the results so far, and this would come out again.

In late 1988, the Working Group was also requested to study the Electricity Plan 1989 – 1998 of the Cooperating Electricity Producers (SEP) in The Netherlands. This prompted the Group to reconsider the strategies for the future. Every two years, the electricity companies submit a joint ten-years investment plan to the Ministry of Economic Affairs for final approval. The government still avoided the decision to reinvest in nuclear energy, and the electricity companies felt the need to keep the nuclear option open as long as possible. Hence, the Electricity Plan 1989 – 1998 covered the increasing electricity demand and the replacement of obsolete plants, partially by the investment in gas-fuelled units, and partially by electricity imports from the neighbouring countries. A strategy with large-scale imports (whereby nuclear energy is in principle accepted) was totally unforeseen in the National Energy Outlook 1987. Anyway, it was an escape route out of the stalemate created by the fruitless energy debate during the nineteen eighties.

We also discussed the results of our experiments with members of the General Energy Council in The Netherlands. The Council, established in 1976 by the Secretary of State for Economic Affairs to advise him on energy-related policy matters, consisted of twenty members from various sectors of our society. In 1989, their predominant problem was to formulate an advice on new investments in nuclear energy. There was a general feeling in the Council that the "yes" or "no" decision could not longer be postponed and that the electricity companies should be enabled to put the next ten-years plan on a more solid basis. Most Council members, however, rejected our suggestion to use a formalized decision analysis because they did not want to arrive at a group compromise solution. They agreed that a unanimous advice would be stronger than a set of divided recommendations, but they insisted that the final decision should be taken by politicians confronted with the rich variety of views and opinions in our society. This argument might have concealed the Council's inability to reconcile the opposite standpoints within its own bosom. Anyway, shortly thereafter the Council submitted a majority advice in favour of nuclear energy, as well as a minority advice advocating a natural-gas strategy to satisfy the base-load electricity demand. This might have been expected after the experiment in the Working Group Integral Energy Scenarios.

In the political arena the alternative strategy based upon electricity imports from neighbouring countries emerged as the preferred strategy. The choice was easily accepted, even by environmental groups, although this strategy clearly supported the nuclear strategies of the neighbours. In the mid-nineties the imports covered 15% of the electricity demand (France 7.5%, Germany 6%, Belgium 1.5%). The nuclear power plant in Dodewaard was closed in 1997. The closure of the plant in Borssele is foreseen in 2003 or 2004. That will be the end of our domestic nuclear capacity.

Nothing has really been solved, however. Safety, interpreted here as the absence of nuclear risk, cannot be guaranteed because our country is surrounded by nuclear power plants. Some of them, the Belgian plants in the neighbourhood of Antwerpen, can actually be seen from Dutch territory. The Energy Report 1994 briefly mentioned that Norway and The Netherlands had roughly 37% each of the West-European natural-gas reserves, and the National Energy Outlook 1995 - 2020 signalized that our country became increasingly dependent on natural gas. The national reserves would largely be depleted in 2020, so that electricity generation thereafter would predominantly depend on natural-gas imports from Russia. Nuclear energy, however, was not an alternative option in the National Energy Outlook 1995 – 2020, despite our dependence on fossil fuels and our commitment to reduce the CO_2 emissions.

Remarks on the methodology. The scale sensitivity of the terminal scores did not really surprise the users in the full-scale experiment because many of them had a strong background in science and technology. Rank reversals due to scale variations had an adverse effect, however. This prompted us to reconsider the algorithmic steps in the AHP. The results may be found in Chapter 4. Furthermore, we initially expected that we would also have to work with plausible scenario probabilities in order to judge the performance of the strategies. It turned out to be unnecessary, but otherwise we could have started from the approach in Section 3.8. Finally, the experiment demonstrated, in a highly convincing manner, that MCDA cannot force the members of a group to accept a compromise solution.

8.6 POLARIZATION OF THE ENERGY DEBATE

Since the early seventies several countries have been the scene of heated energy debates. The critical issues were nuclear power and the storage of nuclear waste. The advocates and the opponents of nuclear power, starting from almost irreconcilable viewpoints, were unable to find a compromise, even when the environmental damage of fossil fuels emerged (acid rain, dying forests). Hence, the energy debates usually ended in a stalemate like in The Netherlands.

Because nuclear technology is highly science-based, scientists were involved in the debates from the beginning. Remarkably enough, they did not appear to be the objective and passionless truth-seekers as they were supposed to be (see also Mitroff, 1974). On the contrary, their research was driven by their personalities. In the political arena they were unable to provide impartial and unanimous advice. Their recommendations largely followed, not from a cool and detached analysis, but from their attitude, that is, from their learned predisposition to respond in a consistently favourable or unfavourable manner with respect to certain items (Fishbein and Ajzen, 1975). In other words, their recommendations were dependent on their fabric, shaped by past experiences, on their manner to filter and to process the available information, on their choice of a viewpoint to see the items in perspective (Chapter 5).

The question of how the attitude of the scientists in the energy debate can be explained in terms of past experiences was one of the subjects in a study by Arts (see Arts et al., 1994). Following the mode of operation of Mitroff (1974) in the USA and Nowotny (1979) in Austria, she submitted a questionnaire to 141 scientists who had been involved in the debate in our country, that is, to scientists of the type which we also encountered in advisory bodies like the Energy Research Council and the General Energy Council (active researchers with an outstanding reputation, and high-ranking research managers). The respondents, 86 in total, specified their viewpoints on the nuclear issue and also their political preferences, their involvement in matters of religion and the environment, their age, and their rank-order position in the parental home. Among the respondents there were 58 advocates (67%) and 28 opponents (33%) of nuclear energy, which enabled Arts to sketch a relationship between the standpoints and the factors just mentioned.

The advocates usually have the confidence that our political and social systems will be strong enough to keep the nuclear technology and the storage of nuclear waste under control. Problems may unexpectedly arise, but science and technology will be able to find appropriate solutions. Nuclear energy is the unique large-scale alternative for fossil fuels, so that it is essential for the diversification of the energy system and for environmental protection. The opponents, however, don't have so much confidence in our political and social systems, and they repeatedly point at the long-term risk of nuclear waste. The economic benefits of nuclear energy are doubtful because the calculations are usually based on weak assumptions, with the research and development costs left out of consideration. Many opponents think that our society needs a new lifestyle, the basis for a careful utilization of renewable energy.

Table 8.5 shows the results of the study by Arts et al. (1994). It will immediately be clear that there are indeed striking differences between the advocates and the opponents. The political preferences of the advocates are mostly at the right-hand side (Liberals, Christian Democrats, to some extent the Democrats 66), whereas the sympathy of the opponents is strongly at the left-hand side (Labour, Green Left). The opponents are much more active in political parties. The advocates have stronger links with religious denominations, and they followed a smoother career path, mostly a university career only. Many advocates are first-born children in the parental family, and almost all of them have been born before 1948 so that they were at least 25 years old when the energy debate started in our country (1973), shortly after the appearance of the reports of the Club of Rome (1972). Many of them strongly remember the Second World War. Therefore they usually assign high weights to economic criteria such as security of the energy supply and support of the national economy. Younger generations assign high weights to other criteria such as safety and environmental protection.

It is a matter of course that the study has far-reaching implications for MCDA. Our approach for group decision making (Chapter 6) is based upon the assumption that the group has a common system of preferences and values so that MCDA may be used to fathom the system and to propose a compromise solution which is properly anchored in it. In the advisory bodies where we carried out our experiments, however, either such a system did not exist, or MCDA was not powerful enough to fathom it in a proper manner.

On the other hand, the members of the advisory bodies were not responsible for the national energy supply. They were not the real decision makers, although their recommendations were widely used to affect the final decisions, and they were not confronted with the issue that the external authority of a group depends on the internal discipline (the bimodal symmetry, see Section 6.1). So, it is possible that a common system of preferences and values, hidden at a deeper level of understanding, only emerges under high pressure, when the group has to make a crucial decision with far-reaching consequences.

Table 8.5 *Political preferences, involvement in political and religious activities, position in the parental family, and age of the advocates and the opponents of nuclear energy for electricity generation. Study by F. Arts et al.* (1994).

	Advocates of nuclear energy $N = 58$		Opponents of nuclear energy $N = 28$	
Liberals	13	22%	1	4%
Christian Democrats	18	31%	0	0%
Democrats '66	12	21%	4	14%
Labour	4	7%	11	39%
Green Left	1	2%	9	32%
Members of a political party	13	22%	13	46%
Communicant church members	27	47%	3	11%
Smooth career path	34	59%	9	32%
First-born in parental family	31	53%	5	18%
Born before 1948	52	90%	15	54%

REFERENCES TO CHAPTER 8

1. Arts, F., Ruiter, W. de, and Klaassen, R., "Belief in Nuclear Energy: Attitudes and Opinions of Dutch Nuclear Experts" (in Dutch). *Milieu, Tijdschrift voor Milieukunde* 9, 163 – 173, 1994.

2. Fishbein, M., and Ajzen, I., *"Belief, Attitude, Intention, Behavior"*. Addison-Wesley, Reading, Mass., 1975.

3. Geus, A.P. de, "Planning as Learning". *Harvard Business Review,* March-April 1988, pp. 70 – 74.

4. Leemhuis, J.P., "Using Scenarios to Develop Strategies". *Long Range Planning* 18, 30 – 37, 1985.

5. Lootsma, F.A., Boonekamp, P.G.M., Cooke, R.M., and Oostvoorn, F. van, "Choice of a Long-Term Strategy for the National Electricity Supply via Scenario Analysis and Multi-Criteria Analysis". *European Journal of Operational Research* 48, 189 – 203, 1990.

6. Mintzberg. H., *"The Rise and Fall of Strategic Planning"*. Prentice Hall, Englewood Cliffs, NJ, 1994.

7. Mitroff, I.I., *"The Subjective Side of Science, a Philosophical Inquiry into the Psychology of the Apollo Moon Scientists"*. Elsevier, North-Holland, Amsterdam, 1974.

8. The National Energy Outlook 1987 (in Dutch). Report ESC-42, Energy Study Centre, 1755 ZG Petten, The Netherlands, 1987.

9. Nowotny, H., *"Kernenergie: Gefahr oder Notwendigkeit. Anatomie eines Konflikts"*. Suhrkamp, Frankfurt am Main, 1979.

10. The Energy Report 1993 (in Dutch). Energy Study Centre, 1755 ZG Petten, The Netherlands, 1994.

11. The Energy Report 1994 (in Dutch). Energy Study Centre, 1755 ZG Petten, The Netherlands, 1995.

12. The National Energy Outlook 1995 – 2020 (in Dutch). Report ECN-C-97-081, Energy Study Centre, 1755 ZG Petten, The Netherlands, 1997.

CHAPTER 9

CONFLICT ANALYSIS AND NEGOTIATIONS

Pairwise-comparison methods can effectively be used to generate ratio information about the benefits and the costs of possible concessions which may be exchanged between two parties in mutual conflict. The basic step is the evaluation of a one-to-one deal where each party offers exactly one concession. The trade-off between benefits and costs is judged in verbal terms which are subsequently converted into numerical values on a geometric scale. The information to be used by a mediator between the two parties appears to be scale-independent. The approach, originally developed for a conflict between two parties, can easily be extended to situations where three or more parties have conflicting interests.

9.1 PAIRWISE COMPARISON OF CONCESSIONS

We are concerned, first, with a conflict between two parties, each being able to offer one or more concessions to the other in order to reduce the tension or to resolve the conflict. We suppose that concessions cannot be made unilaterally but rather in mutual exchange. Thus, each party has to consider the subjective value of every possible deal, that is, each party has to estimate the costs of its own concession in the deal versus the benefits of the concession offered by the adversary.

We shall also be assuming that each party has a representative who thoroughly knows the feelings of the members. In the basic step of the evaluation procedure, a representative is asked to compare two concessions: one made by the adversary and one by his/her own party. Thus, taking $B_i > 0$ to denote the benefit of the adversary's i-th concession and $C_j > 0$ the cost of his/her own party's j-th concession, we ask the representative to estimate the trade-off B_i/C_j of the one-to-one deal (i; j). Let us take the symbol r_{ij} to denote the estimate. The proposed method would hardly be acceptable if the representative would have to estimate the trade-off numerically. That is the reason why he/she is asked to express the intensity of his/her comparative judgement in verbal terms. Thus, believing that $B_i \approx C_j$ the representative may declare his/her party to be indifferent between the two concessions. His/her party may have a weak, definite, or strong liking for the deal (i; j) accordingly as B_i exceeds C_j, and it may have a weak, definite, or strong aversion for the deal accordingly as B_i is exceeded by C_j. In fact, the representative is merely asked to choose one of the gradations displayed in Table 9.1, in order to express how strongly his/her party likes or dislikes the deal.

Table 9.1 *Assessment of the one-to-one deal (i; j). The representative is requested to choose one of the gradations designated by the gradation index, in order to express the strength of his/her comparative judgement of the concessions in the deal.*

Strength	Gradation of judgement	Gradation index δ_{ij}
yes!!	strong liking for the deal	6
yes!	definite liking for the deal	4
yes	weak liking for the deal	2
indifference	indifference	0
no	weak aversion for the deal	-2
no!	definite aversion for the deal	-4
no!!	strong aversion for the deal	-6

Following the analysis in Chapter 4, we shall be assuming that graded comparative judgement of concessions can also be put on a scale with geometric progression. Thus, we write the numerical value r_{ij} which designates how strongly the representative's party likes or dislikes the deal (i; j) in the form

$$r_{ij} = 2^{\gamma\delta_{ij}}, \tag{9.1}$$

where δ_{ij} stands for the gradation index as shown in Table 9.1 (see also Table 4.3). The scale parameter γ is not crucial in the analysis. In the next section we will show that we do not have to assign a numerical value to it. Obviously, the benefits and the costs of the concessions are not really conceived here in monetary terms. In a bitter conflict, concessions may have symbolic or emotional values which largely exceed their economic values. We are therefore concerned here with the concessions as they are emotionally perceived by the two parties within the framework of the actual conflict. It is difficult, if not impossible, to frame the representative's comparative judgement. Hence, we did not introduce something like the range of acceptable performance here.

Even when the full matrix $R = \{r_{ij}\}$ is given, the benefits B_i and the costs C_j cannot individually be estimated, but the ratio information is sufficient to provide improved trade-off estimates t_{ij} of the respective trade-offs B_i/C_j. The full matrix $T = \{t_{ij}\}$ will enable the representative to identify possible inconsistencies in his/her judgement and to discuss the emerging pattern of acceptable deals with the members of his/her party. That will be explained in Section 9.3. Eventually, there will be two matrices of improved trade-off estimates, one for each party. They can be handed over to a mediator. The information in the two matrices will hopefully enable him/her to identify one-to-one deals which are in principle acceptable for both parties.

9.2 IMPROVED TRADE-OFF ESTIMATES

It is possible to smooth the representative's comparative judgement via the method of logarithmic regression. In order to estimate the benefits B_i and the costs C_j, given the original trade-off estimates r_{ij}, we minimize the sum of squares

$$\sum_{i=1}^{m}\sum_{j=1}^{n}\left({}^2\log r_{ij} - {}^2\log b_i + {}^2\log c_j\right)^2 \tag{9.2}$$

as a function of the parameters $b_1,\ldots, b_m, c_1,\ldots, c_n$. The symbols m and n stand for the number of concessions in the hands of the representative's adversary and his/her own party respectively. With the additional notation $\rho_{ij} = {}^2\log r_{ij} = \gamma\delta_{ij}$, $\beta_i = {}^2\log b_i$, and $\gamma_j = {}^2\log c_j$, we can equivalently try to approximate ${}^2\log B_i$ and ${}^2\log C_j$ by minimizing

$$\sum_{i=1}^{m}\sum_{j=1}^{n}\left(\rho_{ij} - \beta_i + \gamma_j\right)^2 \tag{9.3}$$

as a function of the parameters $\beta_1,\ldots, \beta_m, \gamma_1,\ldots, \gamma_n$. The associated normal equations are given by

$$\sum_{j=1}^{n}\left(\rho_{ij} - \beta_i + \gamma_j\right) = 0,$$

$$\sum_{i=1}^{m}\left(\rho_{ij} - \beta_i + \gamma_j\right) = 0.$$

These equations are dependent: there is an additive degree of freedom, to be denoted by ζ. Using the normal equations we can write

$$\beta_i = \frac{1}{n}\sum_{j=1}^{n}\rho_{ij} - \frac{1}{n}\sum_{j=1}^{n}\gamma_j,$$

$$\gamma_j = -\frac{1}{m}\sum_{i=1}^{m}\rho_{ij} + \frac{1}{m}\sum_{i=1}^{m}\beta_i.$$

In order to obtain a particular solution of the normal equations we can arbitrarily set the sum of the γ_j to zero, whence

$$\hat{\beta}_i = \frac{1}{n}\sum_{j=1}^{n}\rho_{ij},$$

$$\hat{\gamma}_j = -\frac{1}{m}\sum_{i=1}^{m}\rho_{ij} + \frac{1}{mn}\sum_{i=1}^{m}\sum_{j=1}^{n}\rho_{ij}.$$

A general minimum solution of (9.3) can be written as

$$\hat{\beta}_1 + \varsigma, \ldots\ldots, \hat{\beta}_m + \varsigma, \hat{\gamma}_1 + \varsigma, \ldots, \hat{\gamma}_n + \varsigma. \tag{9.4}$$

Hence, ${}^2\log B_i$ and ${}^2\log C_j$ cannot individually be estimated, but the difference

$${}^2\log B_i - {}^2\log C_j$$

can uniquely be approximated by

$$\left(\hat{\beta}_i + \varsigma\right) - \left(\hat{\gamma}_j + \varsigma\right) = \hat{\beta}_i - \hat{\gamma}_j = \frac{1}{m}\sum_{i=1}^{m}\rho_{ij} + \frac{1}{n}\sum_{j=1}^{n}\rho_{ij} - \frac{1}{mn}\sum_{i=1}^{m}\sum_{j=1}^{n}\rho_{ij}. \tag{9.5}$$

The improved trade-off estimate t_{ij} of the ratio B_i/C_j is accordingly given by

$$t_{ij} = 2^{\hat{\beta}_i - \hat{\gamma}_j} = \frac{\sqrt[m]{\prod_{i=1}^{m} r_{ij}} \times \sqrt[n]{\prod_{j=1}^{n} r_{ij}}}{\sqrt[mn]{\prod_{i=1}^{m}\prod_{j=1}^{n} r_{ij}}}. \tag{9.6}$$

In order to see that some sort of smoothing has indeed been achieved, we write the original trade-off estimates in the form

$$r_{ij} = \frac{B_i}{C_j}\left(1 + \varepsilon_{ij}\right),$$

where ε_{ij} denotes the relative error in r_{ij}. The factors in (9.6) can now approximately be written as

$$\sqrt[m]{\prod_{i=1}^{m} r_{ij}} \approx \frac{\sqrt[m]{\prod_{i=1}^{m} B_i}}{C_j}\left(1+\frac{1}{m}\sum_{i=1}^{m}\varepsilon_{ij}\right),$$

$$\sqrt[n]{\prod_{j=1}^{n} r_{ij}} \approx \frac{B_i}{\sqrt[n]{\prod_{j=1}^{n} C_j}}\left(1+\frac{1}{n}\sum_{j=1}^{n}\varepsilon_{ij}\right),$$

$$\sqrt[nm]{\prod_{i=1}^{m}\prod_{j=1}^{n} r_{ij}} \approx \frac{\sqrt[m]{\prod_{i=1}^{m} B_i}}{\sqrt[n]{\prod_{j=1}^{n} C_j}}.$$

With these approximations the estimate t_{ij} takes the form

$$t_{ij} \approx \frac{B_i}{C_j}\left(1+\frac{1}{m}\sum_{i=1}^{m}\varepsilon_{ij}\right)\left(1+\frac{1}{n}\sum_{j=1}^{n}\varepsilon_{ij}\right).$$

This implies that the relative errors in the original trade-off estimates have been replaced by average errors in the improved trade-off estimates.

The logarithm of the improved trade-off estimate t_{ij} can also be written as

$$\hat{\beta}_i - \hat{\gamma}_j = \gamma\left(\frac{1}{m}\sum_{i=1}^{m}\delta_{ij} + \frac{1}{n}\sum_{j=1}^{n}\delta_{ij} - \frac{1}{mn}\sum_{i=1}^{m}\sum_{j=1}^{n}\delta_{ij}\right). \tag{9.7}$$

Hence, the scale parameter γ does not affect the sign of ${}^2\log t_{ij}$. The one-to-one deal $(i; j)$ may be accepted (rejected) when the improved trade-off estimate is greater than 1 (smaller than 1) or, equivalently, when the expression between the brackets in (9.7) is positive (negative), regardless of the scale to quantify the gradations of liking or aversion.

The logarithmic-regression model is very flexible. It can also be used when a party is represented by several members. This enables us to use the pairwise comparison of concessions when there is no agreement within a party on the critical issues of the conflict. Various policies and ideas will accordingly be represented in the committee acting on behalf of the party. We let r_{ija} stand for the original trade-off estimate of B_i/C_j given by committee member or actor a, and we take the symbol A_{ij} to denote the set of members who assessed the one-to-one deal $(i; j)$. We now minimize the sum of squares

$$\sum_{i=1}^{m}\sum_{j=1}^{n}\sum_{a\in A_{ij}}\left({}^2\log r_{ija} - {}^2\log b_i + {}^2\log c_j\right)^2 = \sum_{i=1}^{m}\sum_{j=1}^{n}\sum_{a\in A_{ij}}\left(\rho_{ija} - \beta_i + \gamma_j\right)^2. \tag{9.8}$$

The associated system of normal equations does not have an explicit solution such as (9.5). One has to solve the normal equations in order to obtain a general solution of the form (9.4), and this result can immediately be used to find improved trade-off estimates t_{ij} of the ratios B_i/C_j.

Logarithmic regression can also be applied if there is only one representative who does not necessarily assess all possible one-to-one deals. Then the sets A_{ij} have the cardinality zero or one only.

9.3 AN INDUSTRIAL LABOUR DISPUTE

To illustrate matters, we consider an industrial labour dispute where two parties are involved: management and the employees. There are two reasons for the conflict. Strong foreign competition, and henceforth bleak commercial prospects, force management to reduce the labour capacity within a short period of time. The general economic situation, however, is good. This generates nation-wide an upward pressure on wages and salaries. There are various ways to achieve the reduction of the labour capacity, but in an atmosphere of irritation and stress the conflict has been escalating so far that the employees finally decide to go on strike. In the resulting stalemate management could offer the following concessions:

- a reduced number of working hours per week (from 40 to 38),
- two extra holidays per year,
- reduction of the labour capacity via natural turnover only.

The employees have the following concessions in their hands:

- to drop the extra wage claims,
- to accept forced dismissals,
- to end the current strike.

Let us make two remarks before we proceed to an analysis of the conflict. First, the deal whereby the concession "reduction via natural turnover only" is made in exchange of "acceptance of forced dismissals" is a realistic one: both parties leave a dogmatic, unwavering standpoint. Second, dropping a claim is not really a concession if the claim is unjustified or illegal. In the present case, however, wages and salaries are centrally controlled for the complete industrial branch (iron and steel) via tripartite negotiations between the social partners: government, employers, and trade unions. So, in the company under consideration the extra wage claims are firmly based upon the outcome of these negotiations.

Table 9.2 shows the original trade-off estimates, both in verbal terms and in numerical values. For obvious reasons we have only recorded the logarithms $\gamma\delta_{ij}$ of the original trade-off estimates r_{ij}. The third entry in each cell is the logarithm of the improved trade-off estimate t_{ij}. This quantity is simply calculated by adding the arithmetic means of row i and column j and subtracting the total arithmetic mean of the pairwise-comparison matrix. In Table 9.3 the reader finds the viewpoints of the employees recorded in the same way.

Table 9.2 *Industrial labour dispute. Trade-off analysis by the representative of management. Assessment of benefit/cost ratio of concessions of employees versus management's concessions in verbal terms (first entry in each cell), converted into numerical value on a geometric scale (second entry, in logarithmic form), and improved trade-off estimate (third entry, in logarithmic form).*

Concessions of employees	Concessions of management Reduced labour week (40 →38)	 Two extra holidays per year	 Natural turnover only	Arithmetic row means
No extra wage claims	Yes 2γ 2.5γ	Yes! 4γ 5.8γ	Yes!! 6γ 3.8γ	4.0γ
Forced dismissals accepted	Yes! 4γ 3.8γ	Yes!! 6γ 7.1γ	Yes!! 6γ 5.2γ	5.3γ
End current strike	No! -4γ -4.2γ	Yes 2γ -0.9γ	No!! -6γ -2.9γ	-2.7γ
Arithmetic column means	0.7γ	4.0γ	2.0γ	2.2γ

Table 9.3 *Industrial labour dispute. Trade-off analysis by the representative of the employees. Assessment of benefit/cost ratio of management's concessions versus concessions of employees in verbal terms (first entry in each cell), converted into numerical value on a geometric scale (second entry, in logarithmic form), and improved trade-off estimate (third entry, in logarithmic form).*

Concessions of management	Concessions of employees No extra wage claims	 Forced dismissals accepted	 End current strike	Arithmetic row means
Reduced labour week	Indiff 0γ -1.4γ	No! -4γ -3.4γ	No -2γ -1.4γ	-2.0γ
Two extra holidays per year	No! -4γ -2.7γ	No! -4γ -4.7γ	No -2γ -2.7γ	-3.3γ
Natural turnover only	Yes 2γ 1.9γ	Indiff 0γ -0.1γ	Yes 2γ 1.9γ	1.3γ
Arithmetic column means	-0.7γ	-2.7γ	-0.7γ	-1.3γ

Table 9.4 summarizes the information that could be handed over to a mediator. In each cell he/she will find the logarithms of the improved trade-off estimates t_{ij}, one of them given by the representative of management, and the other by the representative of the employees. A positive (negative) sign implies that the party in question likes (dislikes) the one-to-one deal. Opposite signs in a cell designate conflicting viewpoints, two positive (negative) signs indicate that both parties would accept (reject) the one-to-one deal. The scale parameter γ is clearly redundant in the mediator's final analysis of the conflict.

Table 9.4 *Industrial labour dispute. Improved trade-off estimates of one-to-one deal given by management (first entry in a cell) and by the employees (second entry) to a mediator in the conflict.*

Concessions of employees	Concessions of management					
	Reduced labour week		Two extra holidays/year		Natural turnover only	
No extra wage claims	2.5γ	-1.4γ	5.8γ	-2.7γ	3.8γ	1.9γ
Forced dismissals accepted	3.8γ	-3.4γ	7.1γ	-4.7γ	5.2γ	-0.1γ
End current strike	-4.2γ	-1.4γ	-0.9γ	-2.7γ	-2.9γ	1.9γ

Both parties are obviously willing to reduce the tension, management by the promise to use the natural turnover only, provided that the employees drop the extra wage claims. So, this one-to-one deal could be made, and a few weeks later both parties might repeat the exercise just described. Trade-offs change as time goes by! We do not expect that the conflict can be solved in one single step.

9.4 COMPOSITE DEALS

The evaluation of composite deals via trade-off information about one-to-one deals is a non-trivial matter. As a vehicle for discussion we consider the deal $(i, j; k, l)$ where the adversary offers the joint concessions i and j in exchange for the joint concessions k and l. The estimated value of the deal depends on how the representative defines the emotional value. He/she could decide to approximate the ratio of the geometric means of benefits and costs, which is written as

$$\sqrt{\frac{B_i \times B_j}{C_k \times C_l}}, \tag{9.9}$$

but he/she could also decide to approximate the ratio of the arithmetic means or the ratio of the sums of benefits and costs, which is given by

$$\frac{B_i + B_j}{C_k + C_l}. \tag{9.10}$$

As we have seen in the examples of Section 4.5, the choice between ratios of arithmetic means and ratios of geometric means depends on the situation. Since we did not specify the emotional benefits and costs here, the choice is still open. If the decision maker prefers the expression (9.9), for instance, we may use the general solution (9.3) to express the trade-off estimate of the composite deal ($i, j; k, l$) by

$$2^{0.5(\hat{\beta}_i+\hat{\beta}_j-\hat{\gamma}_k-\hat{\gamma}_l)} = \sqrt{t_{ik} \times t_{jl}} = \sqrt{t_{il} \times t_{jk}}\,. \tag{9.11}$$

The question of whether this trade-off estimate is greater or smaller than 1 (accept or reject the deal) does not depend on the scale parameter γ to quantify the judgemental gradations, and the degree of freedom ζ vanishes. If the representative prefers formula (9.10), however, we may use the general solution (9.3) to set the trade-off estimate to

$$\frac{2^{\hat{\beta}_i+\zeta} + 2^{\hat{\beta}_j+\zeta}}{2^{\hat{\gamma}_k+\zeta} + 2^{\hat{\gamma}_l+\zeta}} = \frac{2^{\hat{\beta}_i} + 2^{\hat{\beta}_j}}{2^{\hat{\gamma}_k} + 2^{\hat{\gamma}_l}}. \tag{9.12}$$

In this expression, the scale parameter γ may affect the recommendation to accept or to reject the deal. The degree of freedom ζ vanishes again.

Let us illustrate the results via the evaluation of the composite deal (2, 3; 1, 2) = (forced dismissals accepted, end current strike; reduced labour week, two extra holidays per year) from the viewpoint of management in the industrial dispute of the previous section. The evaluation of the one-to-one deals may be found in Table 9.2. On the basis of formula (9.11) the trade-off estimate of the composite deal is given by

$$2^{0.5(3.8\gamma-0.9\gamma)} = 2^{0.5(7.1\gamma-4.2\gamma)} = 2^{1.45\gamma}.$$

Formula (9.12) yields the trade-off estimate

$$\frac{2^{5.3\gamma} + 2^{-2.7\gamma}}{2^{1.5\gamma} + 2^{-1.8\gamma}} \approx \frac{2^{5.3\gamma}}{2^{1.5\gamma}},$$

at least if the scale parameter γ is in the neighbourhood of 1. These results reveal that the ratio of arithmetic means or the ratio of sums of benefits and costs (9.10) is hazardous. The contributions of some concessions are practically ignored. It seems wiser to use the ratio (9.9) of geometric means of benefits and costs.

Let us also consider the composite deal ($i, j; k$) where the adversary offers the concessions i and j in exchange for the concession k, and let us take the ratio of the sums of benefits and costs

$$\frac{B_i + B_j}{C_k}$$

as a possible candidate to represent the emotional value of the deal. In fact, we do not see any other candidate that could reasonably be used here. It is estimated by

$$\frac{2^{\hat{\beta}_i+\zeta} + 2^{\hat{\beta}_j+\zeta}}{2^{\hat{\gamma}_k+\zeta}} = \frac{2^{\hat{\beta}_i} + 2^{\hat{\beta}_j}}{2^{\hat{\gamma}_k}},$$

an expression which is not affected by the degree of freedom ζ. However, it may easily happen that some concessions are practically ignored in it.

Therefore, we can only suggest the representative to consider the two concessions i and j jointly as a new concession and to estimate its trade-off ratio with respect to the concession k. In doing so, he/she will clearly be concerned with a new one-to-one deal.

This implies that we shall henceforth restrict ourselves to composite deals with an equal number of concessions from both sides. Furthermore, operations such as multiplication and division are more natural here than addition, because we are concerned with ratios of benefits and costs. The appropriate expression to estimate composite deals therefore is the geometric mean of the improved trade-off estimates of the one-to-one deals in it. Hence, the composite deal (1, 2, 3; 1, 2, 3) – which could also be called the total deal because all possible concessions are involved in it – has the estimated value

$$2^{0\ 33(2\ 5+7\ 1-2\ 9)\gamma} = 2^{2\ 2\gamma}$$

for management, and the estimated value

$$2^{0\ 33(-1\ 4-4\ 7+1\ 9)\gamma} = 2^{-1\ 4\gamma}$$

for the employees (see the main diagonals in Tables 9.2 and 9.3). Obviously, the total deal resolving the conflict seems to be acceptable for management only.

9.5 THE CONFLICT IN SOUTH-AFRICA

An earlier attempt at resolving conflicts via pairwise comparison of concessions was presented by Saaty (1988), when he considered the racial conflict in South-Africa. He distinguished two parties, the White Government and the Black Majority, each of them with a number of possible concessions in their hands. The White Government, for instance, could offer

- to release the political prisoners,
- to invite the ANC to participate in a national convention,
- to remove all forms of apartheid and discrimination,
- to grant more autonomy to local governments,
- to grant citizenship and voting rights to all people in the country,
- to improve the conditions in industry and education for all citizens.

The Black Majority could offer

- to assist in the removal of Western pressure, sanctions, and desinvestments,
- to abandon their own commitment to violent actions,
- to stop boycotts and strikes,
- to protect White political rights and investments,
- to agree to power-sharing, immediately or gradually.

Saaty's method to analyze the conflict and to identify the concessions that could be made was different from ours. He proposed to estimate the benefits and the costs of all possible concessions directly via the original AHP. The White Government, for instance, was supposed to compare the Black concessions in pairs under certain performance criteria, to weigh the criteria, and to aggregate the results in order to obtain the terminal scores, the benefits of the Black concessions as viewed by the White party. These scores have a multiplicative degree of freedom, even in the original AHP, so that the estimated benefits would not be unique. The White Government could estimate the costs of their own concessions in a similar way. The estimated costs would also have a multiplicative degree of freedom, however, so that it would be impossible to estimate the benefit/cost ratios of the one-to-one deals thereafter. The rate of exchange would still be unknown! The Black Majority was supposed to evaluate the White concessions as well as their own concessions according to the same two-step procedure, so that their views on the benefit/cost ratios would also remain unknown.

This prompted us to modify the method, so that the representatives of the respective parties could be asked to estimate the benefit/cost ratios directly, within the context of the actual decision problem (Lootsma, 1989).

9.5 DIVORCE AND THE DIVISION OF PROPERTY

The organizational framework for resolving conflicts via pairwise comparison of concessions (two parties with their representatives, and a mediator) is typically found in the division of property just before or after a divorce. The two parties, husband and wife, are each assisted and/or represented by a lawyer. Their proposed division will eventually be accepted or adjusted in court. The judge operates, to some extent, as the mediator between the two parties. He/she ideally overviews the confidential information given by the lawyers.

Although the legislation with respect to marital conditions seems to vary from country to country, we assume that the delicate issue will always be the division of common property, consisting of the items that have been earned during the marital period. So, ignoring the property which was originally brought in by husband and wife, the lawyers will primarily be concerned with the pension rights, the mortgaged house, the actual and/or the future business profits and savings, and they will also consider the level of income of husband and wife in order to propose an alimony. Finally, if there are young children, they usually have to consider the guardianship by one of the two parents.

Various concessions can be made by each of the two parties. Pension rights and/or an alimony may be renounced in exchange for a fair lump sum to be paid from the business profits or from the earnings in previous years. Because the deal financially separates the two partners, it can have a high emotional value. Pension rights and/or an alimony may also be exchanged for the 50% share in the house and the associated mortgage, possibly an attractive deal for the partner who will remain responsible for the children. Since there is a confusing variety of possible deals, it may be helpful for the two partners to make a list of the feasible concessions, to draw a tableau like in the Tables 9.2 and 9.3, to judge some or all possible deals, and to pencil their verbal judgement in the corresponding cells. In doing so, they at least map the emotional minefield.

9.6 NEGOTIATIONS BETWEEN THREE PARTIES

The method for pairwise comparison of concessions has been extended by Wang (1990) to analyze conflicts and/or negotiations between n parties. In what follows, however, we shall concentrate on three parties. This will not only simplify the notation. In an exploratory project with three parties we were confronted with the limitations of human judgement and imagination (Lootsma et al., 1991). The representatives found it difficult to assess three concessions simultaneously and they had to consider many possible deals, so that the method may be too complicated for $n \geq 4$. Success also depends critically on the underlying decision-support system.

In the basic step of the three-party evaluation procedure, a representative is requested to estimate the trade-off ratio B_{ij}/C_j, where B_{ij} stands for the benefit of the i-th and the j-th concession offered by the first and the second adversary respectively, and C_k for the cost of his own party's k-th concession. We take the symbol r_{ijk} to denote the numerical value of the representative's verbal estimate, and we set

$$r_{ijk} = 2^{\gamma\delta_{ijk}},$$

where the integer parameter δ_{ijk} designates the gradation of the representative's verbal judgement (see Table 9.1). As soon as all possible trade-off ratios have been estimated we may smooth the representative's judgement by solving the logarithmic-regression problem of minimizing the sum of squares

$$\sum_{i,j}\sum_{k}\left({}^2\log r_{ijk} - {}^2\log b_{ij} + {}^2\log c_k\right)^2 .$$

Setting $\rho_{ijk} = {}^2\log r_{ijk} = \gamma\delta_{ijk}$, $\beta_{ij} = {}^2\log b_{ij}$, and $\gamma_k = {}^2\log c_k$, we reduce the problem to the least-squares problem of minimizing

$$\sum_{i\,j}\sum_{k}\left(\rho_{ijk} - \beta_{ij} + \gamma_k\right)^2 .$$

There is an additive degree of freedom ζ in the solutions to the associated set of normal equations, and the general solution to the least-squares problem can be written as

$$\hat{\beta}_{ij} + \zeta, \hat{\gamma}_k + \zeta,$$

but the ratio B_{ij}/C_k has the unique improved trade-off estimate

$$t_{ijk} = 2^{\hat{\beta}_{ij} - \hat{\gamma}_k} .$$

The method can be generalized even further. Each party may have several representatives, and they may have different opinions. Taking r_{ijka} to stand for the numerical value assigned to the estimate of B_{ij}/C_k expressed by representative or actor a, we are confronted with the least-squares problem of minimizing

$$\sum_{i,j}\sum_{k}\sum_{a\in A_{ijk}}\left({}^2\log r_{ijka} - {}^2\log b_{ij} + {}^2\log c_k\right)^2$$

where A_{ijk} stands for the set of actors who actually expressed their opinion about the concessions i, j, and k. The associated normal equations will again be linear so that we can easily obtain a solution. Although there is again an additive degree of freedom in it, we can immediately find improved trade-off estimates which are uniquely determined. This was typically the information which we employed in the project of the next section.

9.7 BALANCED CO_2 EMISSION REDUCTIONS

Since the early eighties, worldwide measures against the greenhouse effect and the subsequent global warming are on the political agenda of many international agencies and national governments. Commissioned by the Energy Study Centre (Petten, North-Holland), Sluys explored the potential of the method just sketched in order to fathom the acceptability of multilateral deals aiming at a reduction of CO_2 emissions (see Lootsma, Sluys, and Wang, 1991). The following countries were involved in the experiment:

- The Netherlands, a rich country with a highly developed industrial sector and a high income per capita. The energy system is mainly gas-oriented.

- Poland, a country in economic transition, with an obsolete industrial sector and a coal-oriented energy system. The income per capita is much lower than in The Netherlands.

- Brazil, a country with huge financial problems and a relatively small industrial sector. The income per capita is relatively low. The CO_2 emissions are mainly due to deforestation.

The greenhouse effect is generally felt to be a common responsibility of all countries in the world, so that the above-named countries have common interests, despite the distances separating them. They obviously stand for three economic blocks. The representatives participating in the experiments (three from The Netherlands, six from Poland, and one from Brazil) were members of the energy research establishments working in close contact with the policy makers in their respective countries. They were supposed to understand the political climate in their countries and the boundary conditions under which the politicians have to operate.

Under certain conditions, the population in each of the above countries may accept CO_2 reduction measures. This depends on the perception of the global warming problem and on the efforts made by other countries. Without such measures, in a business-as-usual scenario, the CO_2 emissions of The Netherlands, Poland, and Brazil in the year 2030 would exceed their respective 1991 emissions by 63%, 96%, and 59% (study by the United Nations' International Panel on Climatic Change, see IPCC 1990). In the project we supposed that the three countries have six possible reduction strategies, each leading to CO_2 emissions in 2030 which deviate from their actual 1991 emissions by 40%, 20%, 0%, -15%, -30%, and –50%. In what follows, we shall refer to these percentages as the emission targets or the concessions.

A possible deal among the three countries is a triple of emission targets, one for each country. In principle, there were $6^3 = 216$ possible deals, but since it is unrealistic to ask the representatives to judge all of them, the amount of work had to be reduced. Each representative was therefore confronted with a number of cases only. In each of them he/she had to evaluate the six possible concessions of his own country against a given bid consisting of two concessions, one from each of the other countries (if the bid was unacceptable anyway, the information system recorded a strong aversion for the six corresponding deals). In total, he/she considered 18 cases in order to estimate the position of his/her own country.

Let us now look at the results. Table 9.5 exhibits the top-ten deals for The Netherlands, under the tacit assumption that the improved trade-off estimates, displayed in logarithmic form, may be taken to stand for the acceptability of the corresponding deals. Since the logarithms are all negative, none of the deals is really acceptable. Even the number-one deal is just rejected. Table 9.6, with the top-ten deals for Poland, shows that slight reductions in Poland (-15%) are acceptable if both Brazil and The Netherlands accept more ambitious emission targets (-30% or –50%). Table 9.7, with the top-ten deals for Brazil, shows that a stabilization or a slight reduction of the CO_2 emissions (0% or –15%) in Brazil may be acceptable if Poland and The Netherlands accept significant reductions (-30% or –50%).

Table 9.5 *The top-ten deals for The Netherlands, with logarithms of improved trade-off estimates.*

	Emission targets (concessions)			Logarithms of
Rank order	The Netherlands	Poland	Brazil	improved trade-off estimates
1	-50%	-15%	+40%	-0.2γ
2	-50%	-50%	-15%	-0.2γ
3	-50%	-50%	+40%	-0.6γ
4	-50%	-15%	-15%	-0.8γ
5	-50%	-15%	+20%	-0.9γ
6	-15%	-50%	-15%	-1.0γ
7	-50%	0%	+40%	-1.0γ
8	-15%	-15%	+40%	-1.2γ
9	-30%	-50%	-15%	-1.2γ
10	-30%	-15%	+40%	-1.3γ

Table 9.6 *The top-ten deals for Poland, with logarithms of improved trade-off estimates.*

	Emission targets (concessions)			Logarithms of
Rank order	The Netherlands	Poland	Brazil	improved trade-off estimates
1	-30%	-15%	-50%	4.0γ
2	-30%	0%	-50%	3.4γ
3	-50%	-15%	-50%	3.0γ
4	-30%	-15%	-30%	3.0γ
5	-50%	-15%	-30%	2.6γ
6	-30%	-50%	-50%	2.4γ
7	-30%	-15%	0%	2.3γ
8	-50%	0%	-50%	2.2γ
9	-30%	-30%	-50%	2.2γ
10	-30%	0%	-30%	2.1γ

Table 9.7 *The top-ten deals for Brazil, with logarithms of improved trade-off estimates.*

Rank order	Emission targets (concessions) The Netherlands	Poland	Brazil	Logarithms of improved trade-off estimates
1	-30%	-30%	+20%	4.6γ
2	-30%	-30%	0%	4.2γ
3	-30%	-30%	-15%	3.8γ
4	-30%	-30%	-30%	2.2γ
5	-50%	-50%	+20%	2.0γ
6	-50%	-50%	0%	1.8γ
7	-50%	+30%	+20%	1.8γ
8	-50%	+30%	0%	1.6γ
9	-30%	-50%	+20%	1.6γ
10	-30%	-50%	0%	1.4γ

Table 9.8 shows the top-ten deals ranked on the basis of the smallest of the three acceptabilities (a minimax criterion). The complexity of the global warming problem makes it hard for the countries to form a common opinion on the acceptability of the proposed deals, see the positive and the negative logarithms. Moreover, it seems to be impossible to estimate the economic effects of the reduction strategies. This has repeatedly been confirmed in the literature, see Grübler and Fuyii (1991) as well as Messner and Nakicénovic (1992), for instance. Hence, we could only fathom the willingness of the countries to accept nontrivial measures on the speculative ground that these are required to protect the environment. The top-ten deals in Table 9.8 indicate where a consensus among the three countries may be found. Let us finally note that the representatives of the three countries all had a strong aversion against the deal (40%, 40%, 40%) with the worst possible targets for 2030. This indicates that the global warming problem is considered to be serious. The representatives clearly thought that a global CO_2 reduction would be necessary.

Table 9.8 *The top-ten deals ranked according to the smallest of the three acceptabilities (the logarithms of the improved trade-off estimates).*

Rank	Emission targets (concessions) The Neth.	Poland	Brazil	Acceptability of emission targets in The Neth.	Poland	Brazil
1	-30%	-50%	-15%	-0.6γ	-0.1γ	0.4γ
2	-50%	-30%	0%	-0.7γ	-0.3γ	0.5γ
3	-50%	-50%	-15%	-0.1γ	-0.9γ	0.6γ
4	-30%	-15%	-15%	-0.9γ	0.7γ	-0.3γ
5	-50%	-50%	0%	-1.0γ	-0.2γ	0.9γ
6	-30%	-30%	0%	-1.1γ	0.3γ	2.1γ
7	-50%	-50%	-50%	-1.2γ	0.7γ	-0.7γ
8	-50%	-50%	-30%	-1.2γ	0.5γ	-0.2γ
9	-50%	-15%	0%	-0.9γ	0.6γ	-1.3γ
10	-30%	-50%	0%	-1.4γ	0.4γ	0.7γ

REFERENCES TO CHAPTER 9

1. Grübler, A., and Fuyii, Y., "Inter-Generational and Spatial Equity Issues of Carbon Accounts". *Energy* 16, 1397 – 1416, 1991.

2. IPCC, "Integrated Analysis of Country Case Studies". Report of the US/Japan Expert Group to the Energy and Industry Subgroup of the International Panel on Climatic Change". United Nations, 1990.

3. Lootsma, F.A., "Conflict Resolution via Pairwise Comparison of Concessions". *European Journal of Operational Research* 40, 109 – 116, 1989.

4. Lootsma, F.A., "Comment on the Negotiation and Resolution of the Conflict in South-Africa". *OriON, Journal of the Operational Research Society in South-Africa* 5, 52 – 54, 1989. In the same issue there is a Response by T.L. Saaty (pp. 55 – 57).

5. Lootsma, F.A., Sluijs, J.M., and Wang, S.Y., "Pairwise Comparison of Concessions in Negotiation Processes". *Group Decision and Negotiation* 3, 121 – 131, 1994.

6. Messner, S., and Nakicénovic, N., "A Comparative Assessment of Different Options to Reduce CO_2 Emissions". Working Paper WP-92-27, IIASA, Laxenburg, Austria, 1992.

7. Saaty, T.L., "The Negotiation and Resolution of the Conflict in South-Africa: the AHP". *OriON, Journal of the Operational Research Society in South-Africa* 4, 3 – 25, 1988.

8. Wang, S.Y., "An Approach to Resolve Conflicts by Trade-Off Analysis". *Systems Science and Mathematical Sciences* 3, 1 – 15, 1990.

CHAPTER 10

MULTI-OBJECTIVE LINEAR PROGRAMMING

In Multi-Objective Linear Programming (MOLP) we are concerned with a continuum of alternatives demarcated by a finite number of linear constraints in a finite-dimensional space. Furthermore, there is a finite number of linear objective functions, and a single decision maker or a decision making body. First, we introduce some basic concepts such as efficient (non-dominated) solutions and the dominance cone, and we consider the geometric properties of the efficient set. Next, we discuss several classes of methods for solving the problem. Finally, concentrating on the ideal-point methods, we set the weights of the objective functions via pairwise-comparison methods in order to control the search of an appropriate compromise solution.

10.1 EFFICIENT (NON-DOMINATED) SOLUTIONS

In real-life optimization problems one is usually confronted with several objectives in mutual conflict. A production manager, for instance, who is responsible for the operations in a factory, does not always want to minimize the costs only. For strategic reasons, he/she may also try to minimize the utilization of scarce resources in order to avoid their exhaustion. In structural optimization one usually has to find a proper balance between the minimization of weight and the maximization of performance. In the national energy model of the Dutch economy (Kok and Lootsma, 1985, see also Section 8.2) nine objective functions have been incorporated, reflecting the critical issues in the energy debate such as minimization of costs, minimization of air pollution (SO_2, NO_x, and dust emissions), minimization of net oil imports, minimization of natural gas consumption from domestic fields, and maximization of profits on domestic natural gas.

Goal programming (Ignizio, 1976) is one of the earliest tools for solving a MOLP problem. For each of the objective functions the decision maker sets certain goals or targets which should be satisfied as nearly as possible. In order to remain within the framework of linear programming, one takes the sum of the absolute deviations from the respective goals as a quality index for a feasible solution. Minimization of the sum will, it is hoped, yield a compromise solution which is acceptable for the decision maker.

Other authors delve deeper into MOLP, see Hwang and Yoon (1981), Zeleny (1982), Chankong and Haimes (1983), Osyczka (1984), and Steuer (1986). In fact, MOLP consists of two subfields: (1) the identification of the efficient solutions where the improvement of one of the objective functions cannot be realized without a deterioration of at least one of the remaining objective functions, and (2) the selection of a compromise solution, that is, an efficient solution where the objective-function values are in a proper balance (at least in the subjective opinion of the decision maker). As we will see, MCDA methods can successfully be used in the selection process when we are working within the framework of the ideal-point methods. MCDA enables the decision maker to operate with the vague concept of the relative importance of the objective functions. The approach can mostly be generalized to solve non-linear optimization problems with multiple objective functions as well.

We consider here the general MOLP problem of maximizing the objective functions

$$(c^i)^T x; \; i = 1,\ldots,p, \tag{10.1}$$

subject to the constraints

$$Ax \leq b, x \geq 0, \tag{10.2}$$

where c^i and x are vectors in the n-dimensional Euclidean space E_n, b a vector in E_m, and A a matrix of order $m \times n$. The superscript T stands for transposition.

Let $\bar{x}^i$ denote a feasible solution where the i-th objective function is maximized. Mostly, the single-objective maximum solutions $\bar{x}^1,\ldots,\bar{x}^p$ do not coincide so that we are forced to find a compromise in an efficient solution. By definition, a feasible solution $\tilde{x}$ is referred to as an efficient or non-dominated solution if there is no feasible solution x such that

$$\begin{aligned} &(c^i)^T x \geq (c^i)^T \tilde{x}, \; i = 1,\ldots,p, \\ &(c^k)^T x > (c^k)^T \tilde{x}, \text{ for some } k, \; 1 \leq k \leq p. \end{aligned} \tag{10.3}$$

Obviously, no objective function can be improved by moving away from $\tilde{x}$ into the feasible set without a reduction of at least one of the other objective functions. It will be clear that interior points of the feasible set cannot be efficient. Hence, the efficient set consists of boundary points only. A clear geometric picture of an efficient solution is obtained when we use the so-called dominance cone

$$\Delta = \{x | (c^i)^T x \geq 0; i = 1, \ldots, p\}. \tag{10.4}$$

With this concept, a feasible solution $\tilde{x}$ is efficient if, and only if, there is no other feasible solution x such that $x - \tilde{x} \in \Delta$. Moreover, the feasible solution $\tilde{x}$ is weakly efficient (weakly non-dominated) if there is no feasible solution x such that $x - \tilde{x}$ is in the interior of Δ. This implies that there is no feasible solution x such that

$$(c^i)^T x > (c^i)^T \tilde{x}, \; i = 1, \ldots, p.$$

In other words, it is impossible to improve all objective functions from a weakly efficient solution simultaneously, but it may be possible to improve some objective functions without a reduction of the remaining ones. In the two-dimensional example of Figure 10.1, with a unique and a non-unique maximum solution for the first and the second objective function respectively, the reader will find weakly efficient solutions (any point on the segment *CD*, except *D*) and efficient solutions (any point on the segment *DE*). One only has to move the dominance cone along the boundary of the feasible set in order to verify this.

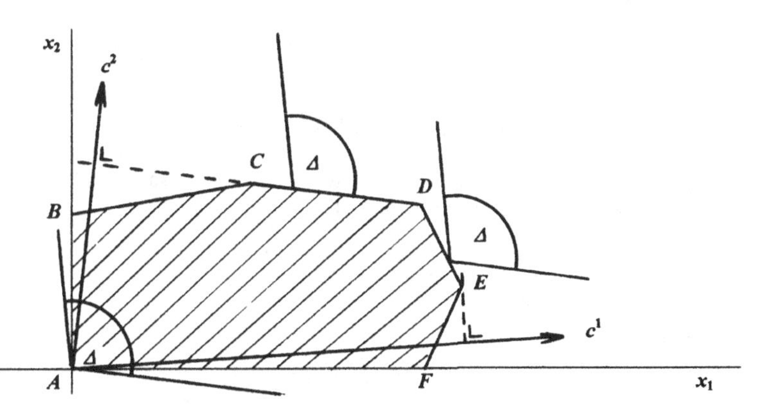

Figure 10.1 *MOLP problem with two objective functions. The second objective function has non-unique maximum solutions. Any point on the segment DE is efficient, any point on CD, with the exception of D, is weakly efficient.*

Frequently, the suggestion is made to solve the MOLP problem by the maximization of a linear combination of the objective functions with positive weight coefficients $\lambda_1, \ldots, \lambda_p$. Thus, we consider the composite objective function

$$\sum_{i=1}^{p} \lambda_i (c^i)^T x = \lambda^T C x, \tag{10.5}$$

where λ stands for the vector with the weight coefficients as the components and C for the matrix with the cost coefficients as the elements. We are now concerned with the problem of maximizing (10.5) subject to the constraints (10.2). The following theorem, which provides a justification for the idea to find at least an efficient solution and hopefully an acceptable compromise solution via the maximization of a positive linear combination of the objective functions, is a key theorem in multi-objective optimization.

Theorem 10.1 (Geoffrion, 1968). A vector $\bar{x}$ in the constraint set (10.2) is an efficient solution of the MOLP problem of maximizing (10.1) over (10.2) if, and only if, there exist positive weight coefficients $\lambda_1, \ldots, \lambda_p$ such that $\bar{x}$ maximizes the composite function (10.5) over the set (10.2).

Proof. If $\bar{x}$ maximizes (10.5) over (10.2), then it is an efficient solution of the original MOLP problem. Otherwise, it would be possible to improve the composite objective function by moving away from $\bar{x}$ into the feasible set. On the other hand, if $\bar{x}$ is an efficient solution of the original MOLP problem, then it is also maximizes the function

$$(c^k)^T x \tag{10.6}$$

subject to the constraints

$$\begin{gathered} (c^i)^T x \geq (c^i)^T \bar{x},\ i = 1,\ldots, p,\ i \neq k, \\ Ax \leq b, x \geq 0, \end{gathered} \tag{10.7}$$

for any $k = 1,\ldots, p$. According to the Kuhn-Tucker necessary conditions for a maximum solution, there exist non-negative multipliers

$$\alpha_1^k, \ldots, \alpha_p^k, u_1^k, \ldots, u_m^k, v_1^k, \ldots, v_n^k$$

such that

$$c^k = -\sum_{\substack{i=1 \\ i \neq k}}^{p} \alpha_i^k c^i + \sum_{i=1}^{m} u_i^k a^i - \sum_{j=1}^{n} v_j^k e^j, \tag{10.8}$$

where a^i stands for the i-th row of the matrix A and e^j for the j-th unit vector. In addition, the multipliers satisfy the complementary slack relations at $\bar{x}$, which are given by

$$u_i^k \left(\sum_{j=1}^{n} a_{ij} \bar{x}_j - b_i \right) = 0,$$

$$v_j^k \bar{x}_j = 0.$$

We rewrite (10.8) as

$$c^k + \sum_{\substack{i=1 \\ i \neq k}}^{p} \alpha_i^k c^i = \sum_{i=1}^{m} u_i^k a^i - \sum_{j=1}^{n} v_j^k e^j .$$

Summing up over k and defining

$$\lambda_i = 1 + \sum_{\substack{k=1 \\ k \neq i}}^{p} \alpha_i^k > 0,$$

$$\bar{u}_i = \sum_{k=1}^{p} u_i^k \geq 0,$$

$$\bar{v}_j = \sum_{k=1}^{p} v_j^k \geq 0,$$

we obtain straightaway the relation

$$\sum_{i=1}^{p} \lambda_i c^i = \sum_{i=1}^{m} \bar{u}_i a^i - \sum_{j=1}^{n} \bar{v}_j e^j .$$

The multipliers $\bar{u}_1,\ldots,\bar{u}_m, \bar{v}_1,\ldots,\bar{v}_n$ also satisfy the complementary slack relations

$$\bar{u}^T (A\bar{x} - b) = 0,$$

$$\bar{v}^T \bar{x} = 0.$$

These results imply that $\bar{x}$ maximizes the composite objective function (10.5) over the feasible set (10.2). Hence, for any efficient solution there are positive weight factors (not necessarily unique) that could have been used to generate the solution in question. This proves Geoffrion's theorem, thereby characterizing the efficient solutions of the original MOLP problem. From the user's point of view, however, nothing has really been solved so far, because he/she mostly has no idea of how to choose the weight coefficients. They may reflect the relative importance of the objective functions, but there is no guarantee that a given set of user-supplied weight coefficients will yield an acceptable compromise solution. In other words, maximization of the composite function (10.5) can be employed to identify efficient solutions, but it does not properly assist the user in the search for a compromise solution.

What happens if we maximize a linear combination of the objective functions with some weight coefficients equal to zero? That depends on the uniqueness of the single-objective maximum solutions. Suppose that we maximize the i-th objective function regardless of the remaining objective functions (which means that we maximize a linear combination of the objective functions with only the i-th weight coefficient different from zero). If the maximum solution is unique, it is also efficient because, whatever we do to improve the other objectives, the i-th objective function will decrease as soon as we move away from the unique maximum solution into the feasible set. The two-dimensional example of Figure 10.1, however, shows that maximization of the linear combination

$$\lambda_1(c^1)^T x + \lambda_2(c^2)^T x, \text{ with } (\lambda_1, \lambda_2) = (0,1),$$

does not necessarily yield an efficient solution if the second objective functions has non-unique maximum solutions. Any point on the line segment CD maximizes the combination, but with the exception of D these points are weakly efficient only, not efficient. When we move along CD to D the first objective function increases without reducing the second one. The points on the segment DE are indeed efficient. The reader can verify this again by moving the dominance cone along the boundary of the feasible set.

It is easier to visualize a MOLP problem, not in the n-dimensional decision space of vectors x, but in the p-dimensional objective space of vectors $z = Cx$. To illustrate matters we consider the MOLP problem of maximizing the objective functions

$$\begin{aligned} z_1 &= 5x_1 - 2x_2 \\ z_2 &= -x_1 + 4x_2 \end{aligned}$$

subject to the constraints

$$\begin{aligned} -x_1 + x_2 &\leq 3 \\ x_1 + x_2 &\leq 8 \\ x_1 &\leq 6 \\ x_2 &\leq 4 \\ x_1, x_2 &\geq 0. \end{aligned}$$

In the two-dimensional decision space the feasible set is the polygon $ABCDEF$ exhibited by Figure 10.2. In the two-dimensional objective space the mapping $z = Cx$ yields the MOLP problem of maximizing z_1 and z_2 over the polygon $A'B'C'D'E'F'$ in Figure 10.3. The efficient solutions can easily be identified here because the dominance cone is given by $\{z \mid z \geq 0\}$. By moving the cone along the boundary of the polygon one obtains that the efficient solutions are on the line segments $B'C'$, $C'D'$, and $D'E'$ in the objective space or, equivalently, on the segments BC, CD, and DE in the original decision space.

10.2 THE CROSS-EFFECT MATRIX

A clear indication of the conflict between the respective objective functions is given by the so-called cross-effect matrix Q with the components

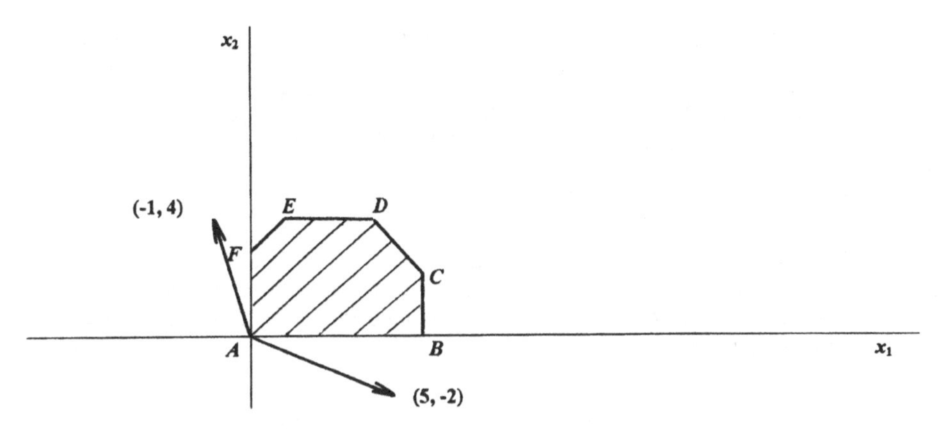

Figure 10.2 *MOLP example in decision space. The feasible set is spanned by the vertices* A = (0, 0), *B* = (6, 0), *C* = (6, 2), *D* = (4, 4), *E* = (1, 4), *F* = (0, 3). *The points on the segments BC, CD, and DE are efficient.*

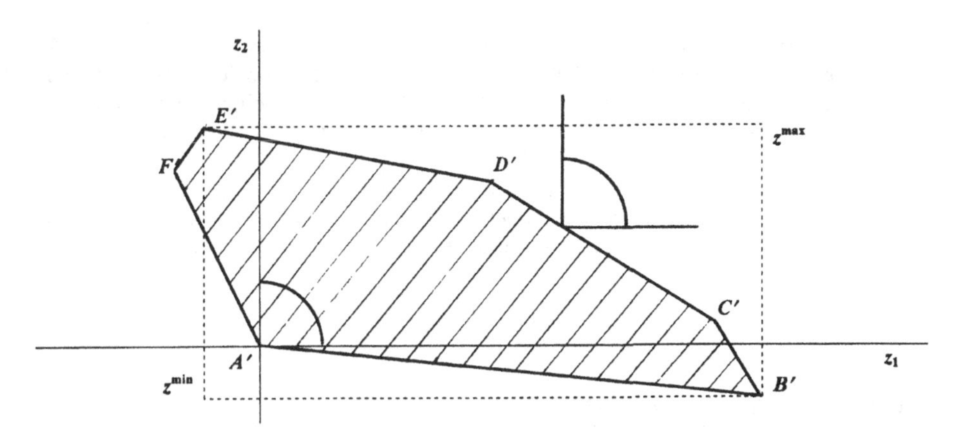

Figure 10.3 *The MOLP example in objective space. The feasible set is spanned by the vertices A′* = (0, 0), *B′* = (30, -6), *C′* = (26, 2), *D′* = (12, 12), *E′* = (-3, 15), *F′* = (-6, 12). *The points on the segments B′C′, C′D′, and D′E′ are efficient. The ideal vector* (30, 15) *and the nadir vector* (-3, -6) *are both unfeasible.*

$$q_{ij} = (c^i)^T \bar{x}^j, \tag{10.9}$$

where $\bar{x}^j$ represents a single-objective maximum solution when the j-th objective function is optimized regardless of the remaining ones. The ideal vector z^{max} is the unique vector with the components

$$z_i^{\max} = (c^i)^T \bar{x}^i, \ i = 1, \ldots, p. \tag{10.10}$$

It is obviously the main diagonal of the cross-effect matrix Q. The nadir vector $z^{\min}$, also referred to as the anti-ideal vector, has as its components the row minima

$$z_i^{\min} = \min_{j=1,\ldots,p} (c^i)^T \bar{x}^j, \ i = 1, \ldots, p, \tag{10.11}$$

which are not necessarily unique because some objective functions may have non-unique maximum solutions. The ideal and the nadir vector are also shown in Figure 10.3. The ideal vector is unfeasible. Otherwise the MOLP problem would have been solved. The nadir vector may be feasible or unfeasible. Usually, we have

$$z_i^{\max} > z_i^{\min}, i = 1, \ldots, p.$$

The components of the nadir vector satisfy the inequalities

$$z_i^{\min} \geq \min_{x \in E} (c^i)^T x, \ i = 1, \ldots, p, \tag{10.12}$$

where E stands for the set of efficient solutions in decision space. Equality in (10.12) cannot be guaranteed, see Steuer (1986). In many applications, however, we use the nadir vector as a vector of pessimistic outcomes for the respective objective functions, in contrast with the ideal vector which we take to represent a vector of optimistic outcomes.

The efficient set can easily be explored by the construction of a new combination of objective functions which is indifferent between the single-objective maximum solutions. The cross-effect matrix plays a crucial role here. We construct the linear combination

$$c^*(x) = \sum_{i=1}^{p} \lambda_i^* (c^i)^T x \tag{10.13}$$

with weight coefficients λ_i^*, $i = 1, \ldots, p$, such that

$$c^*(\bar{x}^1) = c^*(\bar{x}^2) = \cdots = c^*(\bar{x}^p).$$

The weights have to satisfy the equations

$$\lambda_1 q_{11} + \cdots + \lambda_p q_{p1} = 1$$

$$\cdots\cdots\cdots\cdots\cdots\cdots$$

$$\lambda_1 q_{1p} + \cdots + \lambda_p q_{pp} = 1$$

or, equivalently,

$$Q^T \lambda = e, \tag{10.14}$$

where the right-hand side e stands for the p-vector with components 1 (note that the value 1 is not essential; what matters is equality of the right-hand side elements). If we find a positive solution to (10.14) we can further maximize the linear combination (10.13). By Geoffrion's theorem the maximization process will produce an efficient solution.

Applying this procedure to the MOLP problem which is shown in Figure (10.2) we would start off with the single-objective maximum solutions B and E. The linear combination (10.13) would now have contours parallel to BE, and maximization of this function would yield the point D. Starting off with B and D, we would obtain C, but a continuation of the process from B and C would show that further improvements cannot be made. This also happens if we continue from C and D, or from D and E. The efficient set has now fully been explored.

If there are only two objective functions and if the cross-effect matrix is such that

$$q_{11} > q_{12} \text{ and } q_{22} > q_{21},$$

the weights λ_1^* and λ_2^* computed by (10.14) will always be positive.

This is not necessarily true if there are three or more objective functions. We show this via the three-dimensional example of Rietveld (1980) where we are concerned with the maximization of three objective functions

$$\begin{array}{lll} 100x_1 & & \\ 75x_1 & + \ 100x_2 & \\ 40x_1 & & + \ 100x_3 \end{array}$$

subject to the constraints

$$\begin{array}{c} x_1 + x_2 + x_3 = 1 \\ x_1, x_2, x_3 \geq 0. \end{array}$$

A detailed analysis may also be found in Lootsma (1989). The unit vectors e^1, e^2, and e^3 spanning the feasible set are the unique single-objective maximum solutions of the respective objective functions. Hence, they are efficient. The cross-effect matrix is

$$\begin{pmatrix} 100 & 0 & 0 \\ 75 & 100 & 0 \\ 40 & 0 & 100 \end{pmatrix},$$

so that the system (10.14) has the solution $\lambda_1^* = -0.0015$, $\lambda_2^* = \lambda_3^* = 0.01$.

10.3 GEOMETRIC PROPERTIES OF THE EFFICIENT SET

We consider again the original MOLP problem, the maximization of the functions (10.1) subject to the constraints (10.2). As we will see, the characteristic properties of the set E of efficient solutions can be formulated via a few simple theorems. In the Sections 10.4 and 10.5 we shall turn to methods for generating some points of E. It is mostly impossible to generate all of them, particularly if the number of variables is large, but it is also unnecessary because the decision maker is interested in a compromise solution only.

Theorem 10.2 (Yu and Zeleny, 1975). Let x^1 and x^2 be feasible. If x^1 is not efficient, then any point

$$x = \lambda x^1 + (1-\lambda)x^2,\ 0 < \lambda \leq 1,$$

is not efficient.

Proof. There is a feasible point y dominating x^1, so that we have

$$Cy \geq Cx^1,\ Cy \neq Cx^1.$$

Hence, the point $\lambda y + (1-\lambda)x^2$ dominates $\lambda x^1 + (1-\lambda)x^2$ for any $0 < \lambda \leq 1$ because

$$C(\lambda y + (1-\lambda)x^2) = \lambda Cy + (1-\lambda)Cx^2 \geq \lambda Cx^1 + (1-\lambda)Cx^2 = C(\lambda x^1 + (1-\lambda)x^2)$$

and

$$C(\lambda y + (1-\lambda)x^2) \neq C(\lambda x^1 + (1-\lambda)x^2).$$

Corollary. The set of non-efficient solutions is convex.

The set E of efficient solutions is not necessarily convex, see the Figures 10.2 and 10.3.

Theorem 10.3 (Yu and Zeleny, 1975). If the feasible set is bounded so that it is the convex hull of a finite number of vertices $x^1, \ldots\ldots, x^r$, then any efficient solution $\tilde{x}$ is in the convex hull of the efficient vertices.

Proof. Assume the contrary, then there is a non-efficient vertex x^k, $1 \leq k \leq r$, such that

$$\tilde{x} = \sum_{j=1}^{r} \tilde{\alpha}_j x^j,\ \sum_{j=1}^{r} \tilde{\alpha}_j = 1,$$

$$\tilde{\alpha}_j \geq 0, j = 1, \ldots, r, \tilde{\alpha}_k > 0.$$

If $\tilde{\alpha}_k = 1$, then $\tilde{x} = x^k$, and $\tilde{x}$ would not be efficient. Hence, $0 < \tilde{\alpha}_k < 1$. Now,

$$\tilde{x} = \tilde{\alpha}_k x^k + (1-\tilde{\alpha}_k)\sum_{\substack{j=1 \\ j\neq k}}^{r}\left(\frac{\tilde{\alpha}_j}{1-\tilde{\alpha}_k}\right)x^j = \tilde{\alpha}_k x^k + (1-\tilde{\alpha}_k)y.$$

The previous theorem implies, because x^k and y are feasible and x^k is not efficient, that $\tilde{x}$ is not efficient. This contradiction proves the theorem.

Reversely, the Figures 10.2 and 10.3 show that a point in the convex hull of the efficient vertices is not necessarily efficient.

Theorem 10.4 Let $\tilde{x}$ and x^* be feasible solutions with exactly the same number of active constraints. Then $\tilde{x}$ is efficient if, and only if, x^* is efficient.

Proof. This theorem is a direct consequence of Geoffrion's theorem 10.1. Suppose that $\tilde{x}$ is efficient, then there is a positive weight vector $\tilde{\lambda}$ such that the linear combination $\tilde{\lambda}^T Cx$ is maximized over the feasible set at $\tilde{x}$. The necessary and sufficient conditions for optimality in linear programming are that there exist non-negative vectors $\tilde{u}$ and $\tilde{v}$ of multipliers satisfying the relations

$$\tilde{\lambda}^T C = \tilde{u}^T A - \tilde{v},$$

$$\tilde{u}^T(A\tilde{x} - b) = 0,$$

$$\tilde{v}^T\tilde{x} = 0.$$

At the point x^* it must be true that

$$\tilde{u}^T(Ax^* - b) = 0,$$

$$\tilde{v}^T x^* = 0,$$

implying that $\tilde{\lambda}^T Cx$ is also maximized at x^*. This proves the theorem.

10.4 THE CONSTRAINT METHOD WITH TRADE-OFFS

The intuitively appealing idea underlying the constraint method for solving a MOLP problem is to maximize one of the objective functions (the most important one if it can be identified) subject to the additional constraints that the remaining objective functions must be at or above certain lower bounds. Thus, we consider the auxiliary problem of maximizing the k-th objective function

$$(c^k)^T x$$

subject to the constraints

$$(c^i)^T x \geq L_i, i = 1,\ldots, p, i \neq k,$$

$$Ax \leq b, x \geq 0.$$

The L_i stand for lower bounds supplied by the decision maker. Let $\bar{x}^k$ denote a maximum solution of this problem. As usual in multi-objective optimization we have to answer the following questions.

1. Is $\bar{x}^k$ an efficient solution of the original MOLP problem?

2. Can we find lower bounds such that a given arbitrary efficient solution of the original MOLP problem solves the auxiliary problem?

3. Which information can we supply to the decision maker about the quality of $\bar{x}^k$? How do we enable him/her to modify the auxiliary problem if he/she is not satisfied?

Let us start with the first question. Because $\bar{x}^k$ is a maximum solution of the auxiliary problem, we can invoke the Kuhn-Tucker necessary conditions for optimality. There exist non-negative multipliers

$$\alpha_i^k, i = 1,\ldots, p, i \neq k, u_i^k, i = 1,\ldots, m, v_j^k, j = 1,\ldots, n,$$

such that

$$c^k = -\sum_{\substack{i=1 \\ i \neq k}}^{p} \alpha_i^k c^i + \sum_{i=1}^{m} u_i^k a^i - \sum_{j=1}^{n} v_j^k e^j,$$

under the complementary slack relations

$$\alpha_i^k \{(c^i)^T \bar{x}^k - L_i\} = 0, i = 1,\ldots, p, i \neq k,$$

$$u_i^k \{(a^i)^T \bar{x}^k - b_i\} = 0, i = 1,\ldots, m,$$

$$v_j^k \bar{x}_j^k = 0, j = 1,\ldots,n,$$

where a^i denotes the i-th row of A and e^j the j-th unit vector. It is easy to see now that the condition

$$\alpha_i^k > 0, i = 1,\ldots, p, i \neq k,$$

is sufficient to guarantee efficiency of $\bar{x}^k$. The first equation in the above Kuhn-Tucker conditions can be rewritten as

$$c^k + \sum_{\substack{i=1 \\ i \neq k}}^{p} \alpha_i^k c^i = \sum_{i=1}^{m} u_i^k a^i - \sum_{j=1}^{n} v_j^k e^j.$$

Combining this with the complementary slack relations we conclude that $\bar{x}^k$ maximizes the positive linear combination

$$(c^i)^T x + \sum_{\substack{i=1 \\ i \neq k}}^{p} \alpha_i^k (c^i)^T x$$

over the feasible set of the original MOLP problem. Efficiency of $\bar{x}^k$ is now implied by Geoffrion's theorem 10.1.

If some multipliers α_i^k vanish, the point $\bar{x}^k$ may be inefficient, see Figure 10.4.

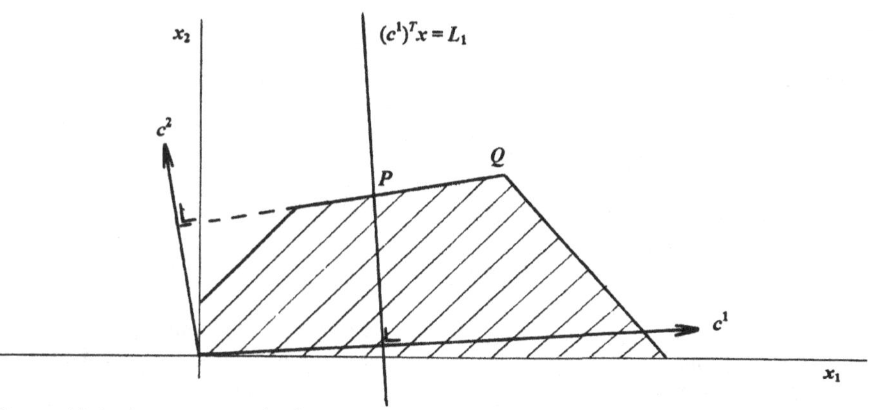

Figure 10.4 *Any point on the line segment PQ maximizes the second objective function subject to the additional constraint that the first objective function is not below the lower bound L_1, but with the exception of Q the points on PQ are weakly efficient only.*

For the second question there is a straight affirmative answer. Let x^* denote an arbitrary efficient solution of the original MOLP problem. If we substitute

$$L_i^* = (c^i)^T x^*, i = 1,\dots,p, i \neq k,$$

into the auxiliary problem, then x^* is a feasible solution. Suppose that it is not a maximum solution of the auxiliary problem, then we would be able to find a point dominating it, and this is a contradiction.

In order to answer the third question, we consider perturbations of the lower bounds in the auxiliary problem, and we investigate the behaviour of the objective function under slight modifications of the bounds. So, in the perturbed auxiliary problem we maximize

$$(c^k)^T x$$

subject to the constraints

$$(c^i)^T x \geq L_i - \varepsilon_i, i = 1,\dots,p, i \neq k,$$

$$Ax \leq b, x \geq 0,$$

with perturbations ε_i, and we take $x^k(\varepsilon)$ to denote a maximum solution.

Now, suppose that the unperturbed auxiliary problem (with all ε_i equal to zero) has positive multipliers

$$\alpha_i^k, i = 1,\dots,p, i \neq k,$$

then the point $\bar{x}^k = x^k(0)$ must be efficient and it must be true that

$$(c^i)^T \bar{x}^k = L_i, i \neq k.$$

For sufficiently small perturbations we must accordingly have

$$(c^i)^T x^k(\varepsilon) = L_i - \varepsilon_i, i \neq k.$$

In a first-order approximation (sensitivity analysis for linear programming) the value of the objective function of the perturbed problem varies linearly with the perturbations in the right-hand side. More precisely, we have

$$(c^k)^T x^k(\varepsilon) - (c^k)^T x^k(0) = \sum_{\substack{i=1 \\ i \neq k}}^{p} \varepsilon_i \alpha_i^k.$$

Combination of the results so far yields the trade-off between the k-th and the i-th objective function which is given by the expression

$$\frac{(c^k)^T\{x^k(\varepsilon)-x^k(0)\}}{(c^i)^T\{x^k(0)-x^k(\varepsilon)\}}=\frac{\sum_{i\neq k}\varepsilon_i\alpha_i^k}{\varepsilon_i}.$$

In particular, if we only perturb the i-th lower bound L_i, the trade-off between the k-th and the i-th objective function is simply given by the multiplier

$$\alpha_i^k.$$

This is important information for a decision maker who is not satisfied with the maximum solution of the unperturbed auxiliary problem. He/she may therefore wish to perturb certain lower bounds. The i-th multiplier shows the increase (decrease) of the k-th objective function per unit of decrease (increase) of the i-th objective function. This is the basic idea of the constraint method. The trade-offs support the decision maker in the adjustment of the lower bounds so that he/she may interactively move towards an acceptable compromise.

10.5 THE IDEAL-POINT METHODS

In practical problems the ideal vector is always unfeasible. Otherwise there would be no conflicts. The nearest feasible solution, however, could be an acceptable compromise solution for the decision maker. As we will see, the acceptability depends largely on the concept of distance from the ideal vector.

We consider the original MOLP problem of maximizing (10.1) subject to the constraints (10.2), and we try to find a feasible solution which minimizes the distance

$$d_\alpha(x)=\sqrt[\alpha]{\sum_{i=1}^{p}\left|z_i^{\max}-(c^i)^Tx\right|^\alpha},$$

where the integer parameter $\alpha, 1\le\alpha\le\infty$, designates the norm in the objective space.

Let us, first, consider the choice of α. The familiar l_1-norm ($\alpha = 1$) and l_2-norm ($\alpha = 2$, the Euclidean norm) are usually not interesting. For $\alpha = 1$, a nearest feasible solution is found by the minimization of

$$d_1(x)=\sum_{i=1}^{p}\{z_i^{\max}-(c^i)^Tx\}$$

over the feasible set, and because the z_i^{max} are constants, we only have to maximize

$$\sum_{i=1}^{p}(c^i)^T x$$

over the feasible set. The solution is clearly efficient because we maximize a positive linear combination of the objective functions (Geoffrion's theorem 10.1), but it is not necessarily unique. When the simplex method is used, the solution will invariably be a vertex of the feasible set. For $\alpha = 2$ the problem of finding a nearest feasible solution reduces to the minimization of a quadratic function subject to linear constraints. There are several successful algorithms for solving this type of problems, but if the original MOLP problem is large, we do not recommend to leave the area of linear programming.

The really interesting ideal-point methods are based upon the l_∞-norm or Chebychev norm. A nearest feasible solution is obtained by the minimization of

$$d_\infty(x) = \max_{i=1,\dots,p}\{z_i^{max} - (c^i)^T x\}$$

over the feasible set. We can equivalently minimize the new variable y subject to the constraints

$$y \geq z_i^{max} - (c^i)^T x, i = 1,\dots,p,$$

$$Ax \leq b, x \geq 0.$$

In the generalizations which we will extensively discuss here we solve the MOLP problem via the minimization of the weighted Chebychev-norm distance

$$\max_{i=1,\dots,p}[w_i\{z_i^{max} - (c^i)^T x\}] \tag{10.15}$$

over the feasible set, with positive weight coefficients w_i. This problem can be rewritten as the minimization of the new variable y subject to the constraints

$$y \geq w_i\{z_i^{max} - (c^i)^T x\}, i = 1,\dots,p,$$

$$Ax \leq b, x \geq 0. \tag{10.16}$$

The idea to minimize the distance function (10.15) by minimizing y over the set (10.16), originally proposed by Benayoun et al. (1971) and also studied by Zeleny (1974, 1977), has been generalized in the reference-point approach of Wierzbicki (1980) underlying the DIDASS program of Lewandowski and Grauer (1982). The method has an attractive feature for the decision maker. If the j-th and the k-th constraint in (10.16) are active at a minimum solution $(\bar{x}, \bar{y})$, then

$$w_j\{z_j^{\max} - (c^j)^T \bar{x}\} = w_k(z_k^{\max} - (c^k)^T \bar{x}\} = \bar{y}, \tag{10.17}$$

which leads to

$$\frac{z_j^{\max} - (c^j)^T \bar{x}}{z_k^{\max} - (c^k)^T \bar{x}} = \frac{w_k}{w_j}.$$

Hence, the deviations of the j-th and the k-th objective-function values from the corresponding ideal values are inversely proportional to the weights. This information supports the decision maker in his/her attempts to find an acceptable compromise solution, although a three-dimensional example is sufficient to show that the minimizing point $\bar{x}$ is not necessarily efficient and not necessarily unique. The example, in principle due to Steuer (1982, 1986), is the maximization of three objective functions x_1, x_2, and x_3 over the convex hull of the unit vectors and a fourth point P in the unit cube. The ideal vector is (1, 1, 1). By an appropriate choice of P and the weights in the distance function (10.15) one obtains that the rectangular contours of the distance function are tangent to the feasible set. Figure 10.5 shows this phenomenon in the cross-section which is spanned by the vertices (0, 0, 0), (0, 0, 1), (1, 1, 0), and (1, 1, 1).

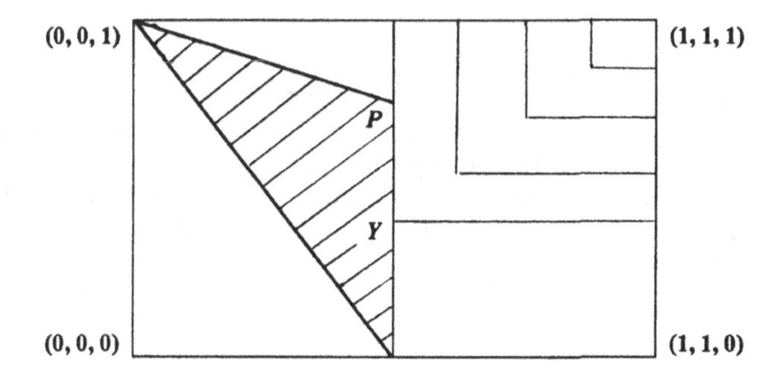

Figure 10.5 *Cross-section through the vertices* (0, 0, 0), (0, 0, 1), (1, 1, 0), *and* (1, 1, 1) *of the unit cube. The shaded area is in the feasible set spanned by the unit vectors and the point P in the cross-section. The figure also shows the contours of the weighted Chebychev-norm distance from the ideal vector* (1, 1, 1). *All points on the line segment PY have the same distance from the ideal vector, but only the point P is efficient (non-dominated).*

In order to avoid the danger of generating feasible solutions which are not efficient, Wierzbicki (1980) proposed to add the term

$$\sum_{i=1}^{m} \varepsilon_i \{z_i^{\max} - (c^i)^T x\}$$

to the distance function (10.15), with sufficiently small positive numbers ε_i. It is the beyond the scope of this chapter, however, to discuss the choice of these numbers.

The decision maker can express his/her reluctance to deviate from the components of the ideal vector by a proper choice of the weights w_i. This feature will extensively be discussed in the next section. It is more convenient to rewrite the weights in the form

$$w_i = \frac{\rho_i}{z_i^{\max} - z_i^{\min}}. \tag{10.18}$$

We cannot in general guarantee that a minimum solution $\bar{x}$ of the distance function (10.15) over the feasible set (10.16) will satisfy the relations (10.17) for all indices j and k, but in practice these relations appear to be satisfied for many j and k. For these indices the ratio of the relative deviations (the deviations from the ideal values as a fraction of the deviations between the ideal and the nadir values) is given by

$$\frac{z_j^{\max} - (c^j)^T \bar{x}}{z_j^{\max} - z_j^{\min}} \Bigg/ \frac{z_k^{\max} - (c^k)^T \bar{x}}{z_k^{\max} - z_k^{\min}} = \frac{\rho_k}{\rho_j}. \tag{10.19}$$

Hence, the decision maker has the approach towards an acceptable compromise solution more or less under control via the choice of the weights. The next section will clarify this. Note that the above ratio (and henceforth the approach) does not depend on the units of performance measurement. If we multiply the objective functions by arbitrary positive constants, the ideal and the nadir values are multiplied by the same values. Hence, the ratio (10.19) of relative deviations from the ideal values is unaffected by a change of the units of measurement.

10.6 OPTIMAL MIX OF ENERGY CARRIERS

In the early eighties, nine objective functions have been implemented in the national energy model which we briefly described in Section 8.2. The project, supported by the Ministry of Economic Affairs (see Kok and Lootsma, 1985), eventually enabled the users to study the following objectives:

1. Minimization of the total costs, in billions (10^9) of guilders 1980.

2. Minimization of SO_2 emissions, in thousands of tons.

3. Minimization of NO_x emissions, in thousands of tons.

4. Minimization of dust emissions, in thousands of tons.

5. Minimization of the exploitation of domestic natural gas, in PJ.

6. Minimization of power generation in nuclear plants, in PJ.

7. Minimization of net oil imports, in PJ.

8. Maximization of profits on domestic natural gas, in billions of guilders 1980.

9. Maximization of power production by combined heat-power units, in PJ.

For a given scenario, based upon assumptions in 1983 and leading to a particular political, technological, and economic situation in the sample year 2000 we obtained the cross-effect matrix of Table 10.1 (see Kok and Lootsma, 1985).

Table 10.1 *Cross-effect matrix of the national energy model with nine objective functions. The matrix generated in* 1984 *reflects the projected situation in the sample year* 2000.

Objectives	1	2	3	4	5	6	7	8	9
1. Total cost	118.3	120.6	122.4	120.6	124.3	120.9	119.4	120.8	119.3
2. SO_2 emission	305.9	212.6	1241.2	253.3	274.6	247.6	218.3	234.6	220.4
3. NO_x emission	419.0	385.7	372.4	392.4	395.5	424.2	397.5	392.9	399.7
4. Dust emission	60.2	51.6	53.8	51.0	79.3	59.6	55.2	51.4	56.3
5. Expl. dom. gas	367.1	459.0	130.3	459.2	33.1	370.8	382.5	485.7	370.8
6. Nuclear power	66.0	66.0	66.0	66.0	66.0	0.0	66.0	66.0	66.0
7. Oil import	1011.0	1015.9	1343.2	1019.5	1411.5	1137.3	968.4	1016.5	972.5
8. Profits on gas	7.5	8.8	4.4	8.8	3.1	7.6	7.8	9.1	7.6
9. Heat-power	70.7	49.8	52.9	49.8	52.9	73.5	68.5	52.9	73.5

The diagonal elements constitute the ideal vector. The maxima of the rows 1 – 7 and the minima of the rows 8 and 9 are the components of the nadir vector. The cross-effect matrix is further illustrated in Figure 10.6, where each vertical bar represents an objective function with the ideal value at the bottom and the nadir value at the top. The j-th level in the i-th bar stands for the value

$$(c^i)^T \bar{x}^j.$$

Obviously, the j-th broken line connects the respective objective-function values at $\bar{x}^j$. A striking message coming out of the cross-effect matrix is that the nuclear power production should be driven up to a predetermined maximum level (66 PJ or 4000 MW) whenever one of the other objective functions is optimized. This level had been imposed by our national government during the energy debate in the nineteen eighties. It rarely happens in real-life projects that a model provides such a clear recommendation.

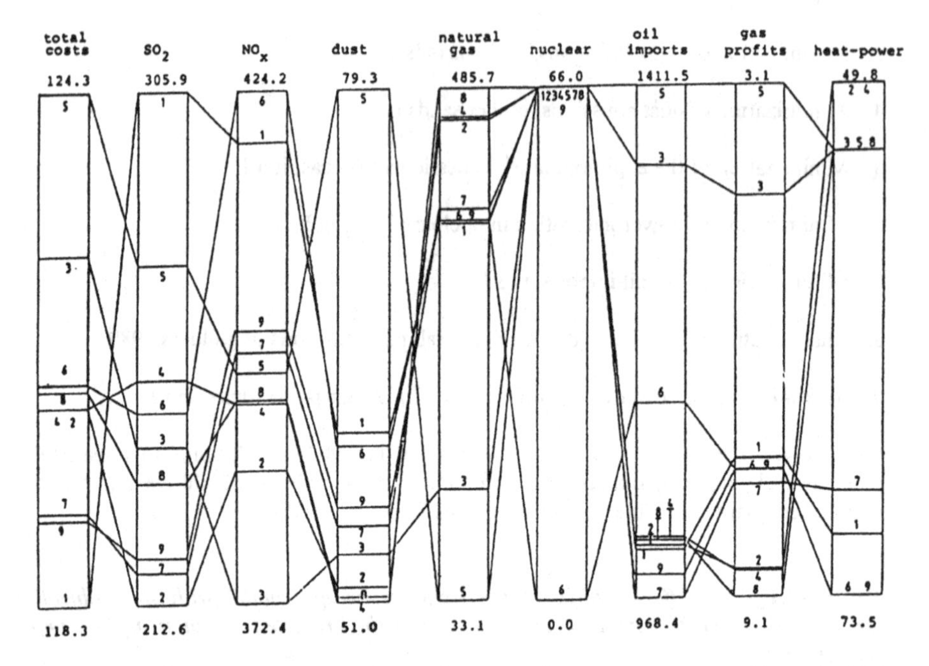

Figure 10.6 *Graphical representation of the cross-effect matrix generated in early* 1984. *Each bar corresponds to an objective function, with the ideal value at the bottom, and the nadir value at the top. The j-th broken line connects the objective-function values at the j-th single-objective maximum solution.*

The experiments to be described in this section consisted of two rounds, like the experiments in Chapter 8. In the first round, six members of the research staff of the Energy Study Centre acted as decision makers. They tested the methodology in an attempt to find a mix of primary energy carriers (coal, oil, natural gas, and uranium) which would constitute an acceptable compromise solution for the long-term energy demand. In the second round, the methodology was used for the same purposes by five individuals representing five organizations with a vital interest in energy matters.

In the first experiment the actors were requested to estimate the weights in (10.15) and (10.18) via pairwise comparisons. In the basic step, we presented the j-th and the k-th objective function to an actor, and we asked him/her to specify the ratio which would be acceptable for the deviations from the ideal vector. In fact, the actor was supposed to estimate the acceptable ratio of ρ_j and ρ_k. The question could be rather precise. We could ask the actor whether a 10% deviation from the ideal value in the direction of the nadir value of the j-th objective function would be equivalent to a similar 10% deviation of the k-th objective function. If the answer would be "yes", if the actor would also be indifferent between 25% deviations in both directions, for instance, and even between 50% deviations in both directions, then the ratio ρ_j/ρ_k could reasonably be estimated by the value of 1. We could also vary the percentages: if the actor would be indifferent between a 10% deviation from the ideal value of the j-th objective function and a 50% deviation from the ideal value of the k-th objective function, then the ratio of the corresponding weights could be set to 5. We would clearly take the inverse ratio of the deviations because a higher weight designates a higher reluctance to deviate from the ideal value. Finally, as soon as all or almost all pairs of objective functions would have been assessed in this manner, we could find the weights via logarithmic regression as described in Chapter 3. Table 10.2 shows the ideal and the nadir vector as well as the deviations which we presented to the actors in order to support their assessment of the objective functions.

Table 10.2 *Objective functions and the corresponding deviations from the ideal value in the direction of the nadir value.*

Objectives	Ideal value	10%	25%	50%	Nadir value
1. Total cost (Dfl 10^9)	118.3	0.6	1.5	3.0	124.3
2. SO_2 emission (1000 tons)	212.6	9.3	23.3	46.6	305.9
3. NO_x emission (1000 tons)	372.4	5.2	13.0	25.9	424.2
4. Dust emission (1000 tons)	51.0	2.8	7.1	14.2	79.3
5. Expl. dom. gas (PJ)	33.1	45.3	113.2	226.3	485.7
6. Nuclear power (PJ)	0.0	6.6	16.5	33.0	66.0
7. Net oil import (PJ)	968.4	44.3	110.8	221.5	1411.5
8. Profits dom. gas (Dfl 10^9)	9.1	-0.6	-1.5	-3.0	3.1
9. Heat-power (PJ)	73.5	-2.4	-5.9	-11.8	49.8

The actors, however, could not easily answer the precise questions submitted to them. They did not pay much attention to the deviations, and they certainly did not compare them. In retrospect, this is not really surprising because the actors were asked to make some awkward comparisons: between reduced SO_2 emissions and increased cost, for instance, or between increased oil imports and increased nuclear power generation. Precise questions of this type were meaningless in the nineteen eighties, and they still are because the effects of SO_2 emissions, the political risk of large oil imports, and the technological risk of nuclear power generation are only vaguely known.

The objectives had emotional or social values, however, and the actors did not hesitate to estimate the relative magnitude of these values, regardless of the information in Table 10.2. In the basic pairwise-comparison step the actor in question was indeed prepared to state whether he/she was indifferent between the two objective functions presented to him/her, or whether one of them was somewhat more, more, much more, or vastly more important than the other. The verbal statements so obtained were converted into numerical values on two different geometric scales, one with the scale-parameter value $\gamma = 0.72$ and one with $\gamma = 1.44$, like in Chapter 4 (given the research results of later years, we would now use the scale-parameter value $\gamma = 0.50$ for the relative importance of the criteria). For each actor and for each value of the scale parameter we eventually had a pairwise-comparison matrix which we further analyzed via logarithmic regression. The resulting normalized weights, for the six actors individually and for the group as a whole, are exhibited in Table 10.3 and Table 10.4. Note that the group weights did not depend on the order of the calculations.

Table 10.3 *Individual weights (six actors) and group weights assigned to the respective objective functions, with verbal comparative judgement converted into numerical values on a geometric scale with* $\gamma = 0.72$.

Objective	1	2	3	4	5	6	Group
1. Total cost	0.069	0.149	0.026	0.136	0.036	0.018	0.062
2. SO_2 emission	0.279	0.149	0.239	0.136	0.182	0.303	0.237
3. NO_x emission	0.200	0.133	0.214	0.169	0.154	0.281	0.221
4. Dust emission	0.129	0.085	0.123	0.056	0.079	0.098	0.105
5. Expl. dom. gas	0.027	0.035	0.214	0.069	0.110	0.026	0.069
6. Nuclear power	0.200	0.259	0.026	0.078	0.146	0.098	0.125
7. Oil import	0.030	0.044	0.063	0.237	0.015	0.036	0.052
8. Profits dom. gas	0.045	0.085	0.032	0.062	0.012	0.026	0.043
9. Heat-power	0.018	0.061	0.063	0.056	0.269	0.115	0.086

Table 10.4 *Individual weights (six actors) and group weights assigned to the respective objective functions, with verbal comparative judgement converted into numerical values on a geometric scale with* $\gamma = 1.44$.

Objective	1	2	3	4	5	6	Group
1. Total cost	0.026	0.144	0.004	0.129	0.006	0.002	0.025
2. SO_2 emission	0.425	0.144	0.329	0.129	0.195	0.436	0.369
3. NO_x emission	0.218	0.115	0.263	0.202	0.139	0.390	0.321
4. Dust emission	0.089	0.071	0.087	0.022	0.037	0.047	0.073
5. Expl. dom. gas	0.004	0.008	0.263	0.034	0.072	0.003	0.031
6. Nuclear power	0.218	0.398	0.004	0.042	0.125	0.047	0.103
7. Oil import	0.005	0.012	0.023	0.393	0.001	0.006	0.018
8. Profits dom. gas	0.011	0.071	0.006	0.027	0.001	0.003	0.012
9. Heat-power	0.002	0.036	0.023	0.022	0.424	0.066	0.048

Eventually, we minimized the objective function (10.15) over the set (10.16), with three different sets of values assigned to the weight coefficients (10.18). The computational task was light: we only had to solve three linear-programming problems with roughly 400 constraints and 500 variables (see Section 8.2). In order to test the model, we first took the weights to be mutually equal (ρ_i = 0.111, i =1,..., 9), and we obtained a solution which will be referred to as Solution A. Thereafter we employed the set of group weights in Table 10.3 and in Table 10.4 to obtain Solution B and Solution C respectively. The corresponding objective-function values are exhibited in Table 10.5. In Figure 10.7 one finds a graphical representation of the three solutions. Solution A is indeed rather flat. Six out of nine objective-function values satisfy the relations (10.17) and (10.19) so that they have the same relative distance from the corresponding ideal and nadir values. The remaining ones (the objectives 4, 8, and 9) are even closer to the respective ideal values. Solution B and Solution C almost coincide, which demonstrates that the scale sensitivity of the approach was low. For practical purposes the discrepancy between the two solutions could be neglected. That was encouraging for our attempts to explore the emotional or social values of the objective functions.

Table 10.5 *Objective-function values at three efficient solutions of the energy model, calculated with three different sets of weights. For Solution A the weights were taken to be equal. For Solution B and Solution C the group weights with scale-parameter values γ = 0.72 and γ = 1.44 have been used. Objective-function values with an asterisk satisfy the relations* (10.17) *and* (10.19).

Objective	Solution A		Solution B		Solution C	
1. Total cost	121.4	*	124.3	*	133.6	*
2. SO_2 emission	261.4	*	237.2	*	228.8	*
3. NO_x emission	399.5	*	387.1	*	382.8	*
4. Dust emission	57.4		57.1		56.0	
5. Expl. domestic gas	270.1	*	130.3		130.3	
6. Nuclear power	34.6	*	33.0	*	41.2	*
7. Oil import	1200.6	*	1426.2		1407.5	
8. Profits domestic gas	6.3		4.4		4.4	
9. Heat-power	63.3		56.3	*	52.9	

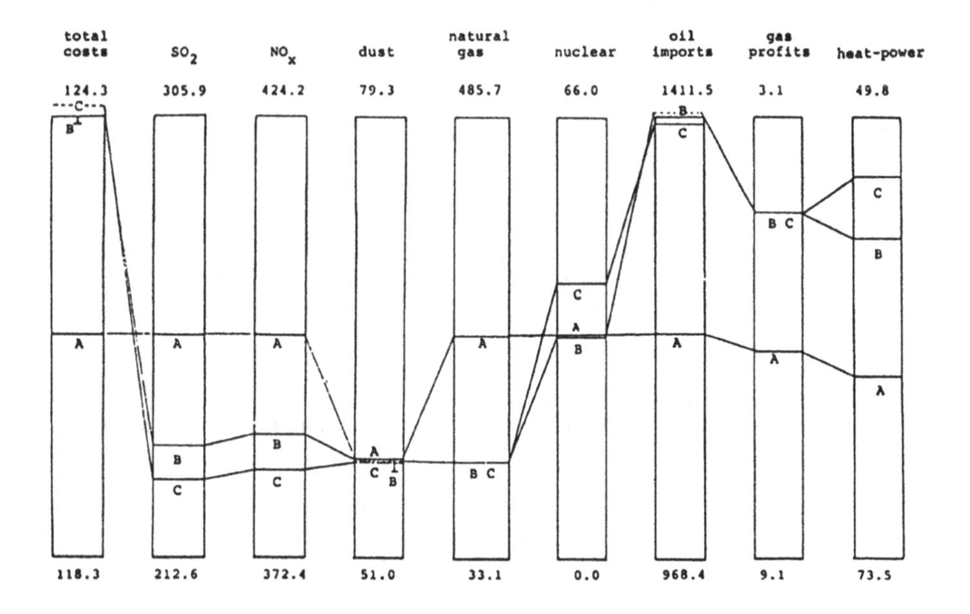

Figure 10.7 *Graphical representation of the three efficient solutions. Solution A is indeed rather flat because the weights were taken to be equal. Solution B and Solution C are close enough to decide that the scale sensitivity of the group weights may be ignored.*

We do not assert that the actors or decision makers should immediately accept the point which minimizes the Chebychev-norm distance from the ideal vector. When they are still unsatisfied with the values of certain objective functions, they can impose lower bounds on them, like in the constraint method, and ask for a repetition of the calculations. This may be the start of an interactive approach, which does not terminate before the actors are fully satisfied.

The second round, conducted and reported by Kok (1986), started with a complete revision of the scenario underlying the original cross-effect matrix in Table 10.1. The projected prices of the primary energy carriers in the sample year 2000 were reduced, the projected demand for secondary energy carriers was lowered, the estimated costs of anti-pollution techniques were changed, etc. In addition, two objectives were dropped: the minimization of dust emissions and the maximization of heat-power combinations. On the other hand, two new objectives were introduced: the minimization of fossil fuels and the maximization of renewable energy. The experiments were carried out with five actors representing, to the best of their knowledge, the viewpoints of the following organizations:

1. The Centre for Energy Conservation, an influential environmental-action group.

2. The Federation of Netherlands Industries, the national employers organization.

3. The Federation of Netherlands Trade Unions, the largest organization of employees.

4. The Ministry of Housing, Physical Planning, and the Environment.

5. The Ministry of Economic Affairs.

The weight elicitation revealed the phenomenon which we also encountered in the first round. The actors did not really pay attention to the cross-effect matrix. In their pairwise comparative judgement they expressed the emotional or social values of the objective functions. Table 10.6 shows the results.

Table 10.6 *Individual weights (five actors) and group weights assigned to the respective objective functions, with verbal comparative judgement converted into numerical values on a geometric scale with* $\gamma = 0.72$.

Objective	1. CEC	2. FNI	3. FNTU	4. HPPE	5. EA	Group
1. Total cost	0.023	0.477	0.058	0.037	0.622	0.158
2. SO_2 emission	0.135	0.126	0.126	0.346	0.105	0.221
3. NO_x emission	0.102	0.112	0.126	0.142	0.105	0.171
4. Fossil fuels	0.108	0.064	0.041	0.042	0.031	0.076
5. Expl. domestic gas	0.053	0.024	0.014	0.019	0.013	0.031
6. Nuclear power	0.347	0.009	0.534	0.091	0.018	0.112
7. Oil import	0.032	0.046	0.024	0.102	0.043	0.064
8. Profits domestic gas	0.030	0.090	0.019	0.022	0.014	0.040
9. Renewable energy	0.169	0.052	0.058	0.198	0.048	0.126

A few comments can be made here. Minimization of total costs and minimization of nuclear power generation appeared to be highly controversial objectives. They were strongly supported by some actors, but they were given very low weights by others. There was a remarkable consensus on the minimization of the air-polluting SO_2 and NO_x emissions. Moreover, all actors assigned low weights to the strategic objectives, the minimization of oil imports and the minimization of domestic natural-gas exploitation. Low priorities were also given to the minimization of fossil-fuel exploitation and to the maximization of profits on domestic natural gas. The question of whether the group weights would produce an acceptable compromise solution was still open. In the calculated solution there was a reasonable nuclear contribution, like in the solution of Figure 10.7. The oil imports appeared to be at the ideal (minimal) value. This was a sharp contrast with the solution of Figure 10.7, where the oil imports were at the nadir (maximal) value.

Several additional experiments, financially supported by the Ministry of Economic Affairs, were carried out in 1985 and 1986 (see Kok, 1986). The number of objective functions was reduced from nine to five, new scenarios and henceforth new cross-effect matrices were developed, etc. The study, which revealed that extensions of the nuclear power-generating capacity would be acceptable for the actors, was favourably received in a wide circle of experts. Kok finished his work shortly before the Chernobyl disaster. Shortly thereafter, however, our national government decided not to resolve the nuclear issue. In Chapter 8 we described how the national policy on energy matters came in a stalemate. Our nuclear-power capacity remained unchanged until 1997. The electricity companies eventually decided to close down the nuclear plant in Dodewaard. The remaining one, the plant in Borssele, will follow the same fate in the beginning of the next century. Roughly 15% of our electricity demand is now covered by imports from the neighbouring countries, which are largely dependent on nuclear power. The energy system as a whole is still heavily dependent on oil imports and on domestic natural gas, despite our international commitment to reduce the CO_2 emission in the next few years.

10.7 MULTI-OBJECTIVE NON-LINEAR PROGRAMMING

Let us finally sketch how the ideas underlying the ideal-point methods can be generalized to solve non-linear programming problems with multiple objectives. We consider the problem of maximizing the objective functions

$$f_i(x), i = 1, \ldots, p,$$

subject to the constraints

$$g_i(x) \geq 0, i = 1, \ldots, m.$$

The problem functions are defined on the n-dimensional vector space E_n, and they are supposed to have continuous second derivatives. Concavity of these functions would be a desirable property as well, but it is difficult to verify. We also assume that the feasible set is closed and bounded, so that each objective function has a maximum solution. For ease of exposition, we assume that the single-objective maximum solutions are unique, and we use the symbol $\bar{x}^i$ to designate the maximum solution of f_i over the feasible set. It will be clear that the success of the ideal-point methods depends critically on the power of global optimization techniques.

The components of the ideal vector z^{max} and the nadir vector z^{min} are given by

$$z_i^{max} = f_i(\bar{x}^i), i = 1,\ldots,p,$$

$$z_i^{min} = \min_{j=1,\ldots,p}[f_i(\bar{x}^j)], i = 1,\ldots,p.$$

By the assumed uniqueness of the single-objective maximum solutions it must be true that the nadir values are also unique and that

$$z_i^{max} > z_i^{min}, i = 1,\ldots,p.$$

With this information one can solve the multi-objective non-linear programming problem via the minimization of the weighted Chebychev-norm distance

$$\max_{i=1,\ldots,p}[w_i\{z_i^{max} - f_i(x)\}]$$

over the feasible set, with positive weight coefficients w_i, $i = 1,\ldots, p$. One can equivalently minimize a new variable y subject to the constraints

$$y \geq w_i\{z_i^{max} - f_i(x)\}, i = 1,\ldots,p,$$

$$g_i(x) \geq 0, i = 1,\ldots,m.$$

Computational experience with this approach in the design of a gearbox has been described by Lootsma et al. (1995). Writing

$$w_i = \frac{\rho_i}{z_i^{max} - z_i^{min}}, i = 1,\ldots,p,$$

one has the search for an acceptable compromise solution more or less under control via the choice of the weights ρ_i. In the feasible point $\bar{x}$ where the distance from the ideal vector is minimized the deviations from the ideal values mostly satisfy the relations

$$w_j\{z_j^{max} - f_j(\bar{x})\} = w_k\{z_k^{max} - f_k(\bar{x})\} = \bar{y},$$

which implies that the deviations from the ideal values in the direction of the nadir values are inversely proportional to the weights. Moreover, the units of measurement do not affect the ratios of the deviations because we can rewrite the ratios in the form

$$\frac{z_j^{\max} - f_j(\bar{x})}{z_j^{\max} - z_j^{\min}} \Bigg/ \frac{z_k^{\max} - f_k(\bar{x})}{z_k^{\max} - z_k^{\min}} = \frac{\rho_k}{\rho_j}.$$

An alternative approach for non-linear programming, reported by Lootsma et al. (1995) and Lootsma (1997), is to maximize the distance from the nadir or anti-ideal vector. Then the problem is to maximize the weighted degrees of satisfaction defined as the weighted geometric mean

$$\prod_{i=1}^{p} \left(\frac{f_i(x) - z_i^{\min}}{z_i^{\max} - z_i^{\min}} \right)^{c_i}$$

subject to the constraints

$$f_i(x) \geq z_i^{\min}, i = 1,\ldots,p,$$

$$g_i(x) \geq 0, i = 1,\ldots,m.$$

The c_i, $i = 1,\ldots, m$, stand for normalized weights assigned to the objective functions. The computational study with the two methods, minimization of the weighted Chebychev-norm distance and maximization of the weighted degrees of satisfaction, demonstrated that the solutions were roughly the same whenever

$$\rho_i = c_i, i = 1,\ldots,p.$$

Obviously, as soon as the ideal vector and the nadir vector are known, the choice of the weights enables the decision maker to home in towards a compromise solution where the ratios of the deviations from the ideal vector are related to what is intuitively known as the relative importance of the objective functions.

REFERENCES TO CHAPTER 10

1. Benayoun, R., Montgolfier, J. de, Tergny, J., and Larichev, O., "Linear Programming with Multiple Objective Functions: STEP Method". *Mathematical Programming* 1, 366 – 375, 1971.

2. Chankong, V., and Haimes, Y.Y., *"Multiobjective Decision Making"*. North-Holland, Amsterdam, 1983.

3. Geoffrion, A.M., "Proper Efficiency and the Theory of Vector Maximization". *Journal of Mathematical Analysis and Applications* 22, 618 – 630, 1968.

4. Hwang, C.L., and Yoon, K., *"Multiple Attribute Decision Making"*. Springer, Berlin, 1981.

5. Ignizio, J.P., *"Goal Programming and Extensions"*. Lexington Books, Lexington, Mass., 1976.

6. Kok, M., *"Conflict Analysis via Multiple Objective Programming, with Experiences in Energy Planning"*. Ph.D. Thesis, Faculty of Mathematics and Informatics, Delft University of Technology, Delft, The Netherlands, 1986.

7. Kok, M., and Lootsma, F.A., "Pairwise-Comparison Methods in Multiple Objective Programming, with Applications in a Long-Term Energy-Planning Model". *European Journal of Operational Research* 22, 44 – 55, 1985.

8. Lewandowski, A., and Grauer, M., "The Reference Point Optimization Approach". In M. Grauer, A. Lewandowski, and A.P. Wierzbicki (eds.), *"Multiobjective and Stochastic Optimization"*. IIASA, Laxenburg, Austria, 1982, pp. 353 – 376.

9. Lootsma, F.A., "Optimization with Multiple Objectives". In M. Iri and K. Tanabe (eds.), *"Mathematical Programming, Recent Developments and Applications"*. KTK Scientific Publishers, Tokyo, 1989, pp. 333 – 364.

10. Lootsma, F.A., *"Fuzzy Logic for Planning and Decision Making"*. Kluwer Academic Publishers, Dordrecht/Boston/London, 1997.

11. Lootsma, F.A., Athan, T.W., and Papalambros, P.Y., "Controlling the Search for a Compromise Solution in Multi-Objective Optimization". *Engineering Optimization* 25, 65 – 81, 1995.

12. Osyczka, A., *"Multicriterion Optimization in Engineering"*. Wiley, New York, 1984.

13. Rietveld, P., "Multiple Objective Decision Methods and Regional Planning". *Studies in Regional Science and Urban Economics* 7, North-Holland, Amsterdam, 1980.

14. Steuer, R.E., "On Sampling the Efficient Set using Weighted Tchebycheff Metrics". In M. Grauer, A. Lewandowski, and A.P. Wierzbicki (eds.), *"Multiobjective and Stochastic Optimization.* IIASA, Laxenburg, Austria, 1982, pp. 335 – 352.

15. Steuer, R.E., *"Multiple Criteria Optimization Theory, Computation, and Application"*. Wiley, New York, 1986.

16. Wierzbicki, A.P., "A Mathematical Basis for Satisficing Decision Making". WP-80-90, IIASA, Laxenburg, Austria, 1980.

17. Yu, P.L. and Zeleny, M., "The Set of All Nondominated Solutions in Linear Cases and a Multicriteria Simplex Method". *Journal of Mathematical Analysis and Applications* 49, 430 – 468, 1975.

18. Zeleny, M., *"Multiple Criteria Decision Making"*. McGraw-Hill, New York, 1982.

CHAPTER 11

MCDA IN THE HANDS OF ITS MASTERS

The MCDA methods, usually developed by specialists in academic research institutes, were mostly designed for decision making in public administration and industrial management. Developed in the ivory tower of cool scientific contemplation, they might be inadequate for the turbulent world of government and business. The specialists with a vivid interest in applications of MCDA therefore had to cross the notorious chasm between theory and practice. Occasionally, however, they also found themselves in the role of decision makers. In such a role, they were in the position to judge the effectiveness of mathematical and computational support for decision making. How did they see the MCDA methods from inside, how did they use their first-hand knowledge about the algorithmic ideas, how did they handle the underlying assumptions, the strengths and the weaknesses, the pitfalls and the benefits of MCDA? How did they work individually with MCDA, how did they apply it in groups, and how did they come to a decision? A thorough study is not yet available. This chapter only sketches our experiences in various situations.

11.1 THE DECISION MAKER AND THE MCDA SPECIALIST

During my academic career I served on several nomination committees entrusted with the task to fill a vacant chair at a university in The Netherlands. The procedures are lengthy (see Section 1.1) and they may take two years, even when the original profile was easily drafted and accepted. On the eve before the decisive meeting of the committee I reread the minutes of previous meetings. I go again through the information supplied by the three or four applicants who are retained in the last round, and finally I summarize the information in a performance tableau in order to use MCDA. It is late in the evening, then, almost midnight. Symbolically speaking, it is eleven fifty five p.m., the moment of truth, not only in the evening but also in the decision process. Nevertheless, MCDA sometimes had an unexpected impact on me. I sometimes drastically changed my opinion, a move which also affected the committee on the next morning. In retrospect it still surprises me how well the analysis brought up the criteria which really mattered in the final decision.

Some committees were prepared to apply MCDA themselves in the final stage of the decision process. Other committees had strong objections for all kinds of well-known reasons: the methods were not sufficiently tested, the analysis might bring up a candidate which would not have been proposed via the traditional style of decision making, it might be easier for the committee to agree on the preferred candidate than on the precise formulation and the weights of the criteria, etc. Some members presumably had a hidden agenda, others thought that decision making is a matter of feeling so that they disliked any quantitative approach.

My experience is different, however. True, the analysis did not increase the efficiency of the committee (the processes were as lengthy as they had always been), but it contributed to the quality of the decision and it increased the satisfaction with the decision process in the last meeting. I also regretfully remember at least one occasion where we did not apply MCDA so that we overlooked some important pieces of information.

Nevertheless, I am unable to explain the benefits of MCDA precisely in the initial phase of project acquisition. I can tell the possible clients, from my own experience, that MCDA improves the formulation of the criteria, that it clarifies how the members weigh the criteria and the alternatives, and that it triggers interesting discussions when there are widely varying opinions in the committee (see also the approach of clients in the case study by French et al., 1998). However, telling the clients that MCDA improves the quality of the decision and/or the decision process could be insulting. It suggests that their style of decision making might have been doubtful so far. It is even more hazardous to tell them that MCDA sometimes suggests the decision makers to change their mind. MCDA does not merely confirm the vague preconceived ideas in a decision-making body.

If I would have strong economic arguments, if I could guarantee that MCDA accelerates the convergence towards consensus, if I could promise a dramatic competitive advantage to the client organization, yes, then I could perhaps easily persuade the possible clients to apply the technique, but I almost never found these arguments. Some industries happily applied MCDA because the regional and the divisional managers, convened at company headquarters for a couple of days, could solve more problems in the given time span than before. In public administration and in universities, however, the pressure of time is lower so that economic arguments do not count so much.

So, this is my dilemma. I did not find the straightforward compelling reasons (improved productivity, significant cost reductions, a competitive advantage, see also Moore, 1991) why the clients should (?) carry out an MCDA study, and I cannot convincingly describe the remaining ones. Is that true for other MCDA specialists as well? And if they have the same difficulty, what is the future of our field?

In the position of a decision maker I am happier with MCDA than in the role of a specialist who has to advertise the technique. As a decision maker I can apply MCDA at any moment of the day or the night; I can do so with or without the support of another specialist; I have first-hand information about the actual decision problem; I am not under the obligation to carry out the analysis until the bitter end; I do not have to conceal anything for myself; I know the strengths and weaknesses of the methods; and I can freely accept or reject the conclusions. To some extent I am a line manager in a university. Hence, I can completely or partially follow the suggestions of the specialists, but I can also ignore their advice. Who are the specialists, however?

They are the analysts of the technostructure as described by Mintzberg (1983). They have no formal authority to make decisions, they are professionals motivated by technical/scientific excellence and they are forced to legitimate themselves by tangible benefits. This is exactly the description of the automation specialists in an industrial staff department where I worked for many years. They used the possible efficiency increase in production planning and inventory control to justify the soaring costs of their automation projects. And indeed, when I am searching for clients to test my ideas and for sponsors to finance my projects, I am also such a person who has to legitimate himself. MCDA is my bread and butter, and I am under the obligation to create a market for MCDA by the performance of the technique.

It cannot be surprising that the decision maker and the specialist should have totally different views on the possible benefits of MCDA. As a decision maker, when I am using MCDA for myself, I am not interested in a possible efficiency increase or in a competitive advantage. It is my concern to understand the importance and the impact of my criteria, and I know that MCDA contributes to such an understanding. As a specialist, however, I have to operate carefully with the arguments to use MCDA. In the first discussion with possible clients, I cannot suggest that MCDA might improve the quality of their decisions, and mostly I cannot use the efficiency increase as an argument at all.

The contrast that I have sketched so far is not entirely fair, however. I depicted myself as an individual decision maker. In general, however, decisions are so frequently made in groups that the specialist usually has to deal with boards, councils, and committees. The members of such a decision-making body sometimes do not even know the criteria and the criterion weights of their colleagues. Let us therefore see how MCDA works in groups where the members are MCDA specialists themselves.

11.2 CHOICE OF WORKING LANGUAGES

Shortly after 1989, a thorny decision problem emerged in the European Working Group "Aide Multicritère à la Décision/Multicriteria Decision Aid". Established in 1975, the Working Group had French as their working language only, but English had gradually been accepted so that the Group (in the early nineties roughly 170 members) became really bilingual. Some 50% of the members, mainly from France, Belgium, and Switzerland, were francophone. The others came from a variety of European countries: Southern Europe, Scandinavia, and The Netherlands. Almost none of them had English as his/her native language. The participants of the workshops, twice per year, were supposed to understand both French and English.

Since the collapse of the Berlin Wall, November 1989, and the unification of Germany, October 1990, the political scene in Europe had thoroughly changed. There was a general feeling that not only German industry but also the German language would deeply penetrate into Central and Eastern Europe. This was not new, however. Until the Second World War, French and German were the common languages in industry, trade, and science on the European continent. The predominant position of the English language dates from 1945. Older generations of scientists, particularly in the smaller European countries, could easily communicate in French and German, but thereafter English became the vehicle for communication in science and technology, even in Europe. More than 95% of all proposals submitted in Brussels for financial support under the ESPRIT, AIM,..., program, for instance, were drafted in English. Nevertheless, many industrial companies realized that their employees should be able to work with a variety of languages in the new European constellation.

In the winter of 1989/1990, I spent two months at LAMSADE, Université de Paris-Dauphine, to work on MCDA. Of course, I discussed the recent collapse of the Berlin Wall and the political revolutions in Central Europe with Bernard Roy, the chairman of the Working Group. We realized that these events could affect the choice of working languages in the Group. At that time the number of German-speaking members was still rather small, so that the question of whether the German language should be adopted as the third working language was not urgent. Nevertheless, it might be wise to prepare the ground for a decision. It was by no means clear that the Working Group, with three working languages, would remain as coherent as it was in the early nineties. It might disintegrate into a number of loosely connected subgroups, each discussing the subjects which were felt to be important in their own region. However, the scientific world was not limited to Europe and North-America, as it was before 1939. Major contributions came from the Far East, where international communication is exclusively in English. Since every scientist was (and is) expected to be fluent in English anyway, the Working Group might therefore avoid any further fragmentation by accepting English as their unique working language. This would have the additional advantage that the Group would be open to visitors from other continents. The disadvantage would be that an English-speaking group, the European Summer Institute Group on Multi-Criteria

Analysis (ESIGMA), already existed, but it was certainly less active than the Working Group Aide Multicritère.

A consequence of the decision to accept English only might be that the Working Group would rapidly grow and loose its attractive features. The workshops had a high frequency, some 40 or 50 members were present at each of them, and there was ample time for informal discussions. The overflow of proposed contributions was transferred to the next workshop. Moreover, the organization of a workshop was rather simple. Larger meetings would considerably increase the task of the organizers. Given such an informal and flexible structure, it could be wise to keep the Group as small as it was and to increase its coherence by taking French as the unique working language.

Thus, there were four alternative options, each to be judged under the criteria of Coherence, Complementarity, and Time for Discussions. It fascinated me that here we had the testing ground for large-scale MCDA with one or two hundred potential decision makers involved in a real-life problem. The choice of working languages would also affect the scientific ambiance, because it would affect the participation by Anglo-Saxon and Asian scientists.

Finally, it would also affect the subjects to be treated at the workshops. There was a French school in MCDA (the constructive approach, see Section 1.3) and an American school (Multi-Attribute Utility Theory, the normative approach), but the cross-fertilization was (and still is) meagre. Well-known Anglo-Saxon textbooks (von Winterfeldt and Edwards, 1986; French, 1988) paid due attention to MAUT and SMART, but they ignored the French school which was amply described in some textbooks on the European continent (Roy, 1985; Schärlig, 1985; Vincke, 1989). None of them presented a thorough description of the AHP although its prescriptive approach, incorporated in the EXPERT CHOICE program, turned out to be much more popular in government and business. It might therefore be more appropriate to refer to the AHP school as the American school (see the author, 1989). Anyway, the proportion of contributions devoted to each of the schools would depend on the choice of working languages.

After my return to Delft, February 1990, I worked out the technical details of a possible large-scale MCDA experiment in cooperation with Leo Rog and Matthijs Kok (Delft University of Technology). It was our intention to prepare and to support the choice via the REMBRANDT system which we also used in other projects. First, we drafted the performance tableau which is shown in Table 11.1. It summarizes the possible consequences of the alternative combinations of working languages under each of the performance criteria just mentioned. Earlier experiments in the Working Group, presented at the workshop in Delft, March 1990, convinced us that it would technically be feasible to submit a decision problem with four alternatives under five criteria to every member and to ask them to return our special questionnaires (see Section 4.3) via the mail. The subsequent computations might produce a compromise solution, a combination of working languages which would be acceptable for the Working Group.

Of course, the search for a compromise solution should be preceded by a thorough discussion of the proposed alternatives and the adopted criteria. In order to test and to illustrate the procedure, however, we submitted the problem to the staff and the visitors of our Department. First, they were requested to judge the performance of the combinations under each of the criteria and to express their judgement by direct rating as in SMART (Chapter 2). Similarly, they were invited to rate the criteria on the basis of relative importance for a scientific working group. Next, we asked them to repeat the evaluation and to judge the alternative combinations in pairs with the help of the familiar questionnaires. The responses were analyzed by means of the Multiplicative AHP (Chapter 3). The experiment was encouraging: nineteen members, originating from The Netherlands (12), Germany (4), China (1), UK (1), and Hungary (1), returned their judgement. Table 11.2 exhibits the calculated criterion weights and terminal scores of the group of respondents. Obviously, English only was the preferred working language. This was also true for 17 participants individually. The leading criteria were Scientific Ambiance and Subjects Treated. Scientific Ambiance had the highest weight for 14 respondents individually.

Table 11.1 *Possible consequences of the choice of working languages: four alternative combinations under five criteria.*

Combination and character of group	English only, international	French only, regional	English + French, status quo	Eng. + Fr. + Ger., continental
Coherence	Group is loosely connected	Monolingual, group is cordial	Bilingual, group is bipolar	Multilingual, group fragmented
Complementary	There is a competing group	No competing group	No competing group	No competing group
Scientific Ambiance	Open to any scientists	No Anglo-Saxon or Asian members	Few Anglo-Saxon or Asian members	Very few Anglo-Saxons or Asians
Subjects Treated	Dominated by American school	Dominated by French school	Mix of French and American school	Mix of French and American school
Time for Discussions	Very limited, very large group	Ample, small group	Rather limited, fairly large group	Limited, large group

Table 11.2 *Group weights of five performance criteria and group terminal scores of four alternative combinations of working languages* (19 *respondents*).

	Direct rating, SMART	Pairwise comparisons, Multiplicative AHP
Coherence	0.143	0.116
Complementarity	0.114	0.088
Scientific Ambiance	0.297	0.403
Subjects Treated	0.278	0.256
Time for Discussions	0.168	0.137
English only	0.432	0.696
French only	0.122	0.025
English + French	0.246	0.148
Eng. + Fr. + Ger.	0.200	0.131

Of course, the calculated weights and scores were not representative for the Working Group. The educational reforms of the nineteen sixties and seventies ruined the position of the French language in our country. The working language in our Department was (and still is) English, although several members had close working relationships with French-speaking colleagues. Technically, however, the experiment confirmed the feasibility of large-scale MCDA. Pairwise comparisons were felt to be complicated and time-consuming, but direct rating appeared to be an attractive tool for experiments with a large group of decision makers.

Since I had taken part in the experiment myself, I had seen my own weights assigned to the criteria and my own terminal scores assigned to the alternative combinations of languages. It puzzled me that I could not easily accept the outcome. The analysis brought up that I also seemed to have a clear preference for English only, just because of my high weights for Scientific Ambiance and Subjects Treated. On the other hand, I liked the access to the French way of life via the Working Group, something that would have been impossible without the current status of the French language in the Group. The social program of the workshops was mostly excellent: in the excursions, receptions, and evening dinners the organizers attended to it that each workshop had a distinct local flavour. From my twelve years experience in a multinational company I also knew that, in order to understand the decision processes in other countries, one has to understand the language first. In reality, I was happy with the current situation in the Group, but the MCDA experiment suggested that I would have to change my mind.

In two consecutive meetings of the Working Group (Marseille, October 1991, and Chania, Greece, March 1992), Leo Rog, Matthijs Kok, and the present author proposed to prepare the choice of working languages via MCDA for the following reasons:

1. It is a real-life problem in the Working Group. We believe that MCDA specialists should occasionally test their own methods on their own decision problems.

2. It provides ample opportunities for a comparative study of various MCDA methods.

3. It is a large-scale problem because there are hundreds of potential decision makers involved in it. Hence, it can also be used to test various decision support methodologies.

There was an absolute silence after our presentation in Chania. Neither the alternatives nor the criteria were rejected, no new criteria were proposed. Only one participant was prepared to take part in the experiment, and we were invited to submit our report for publication in a special issue on multi-criteria decision making (see Lootsma, Rog, and Kok, 1993). One can only guess at the reasons why the Working Group ignored the proposal. Either there was no problem because those normally present at the workshops accepted the status quo, or there was a problem, indeed, but the proper criteria, too delicate to be discussed, were felt to be missing. It was also possible that the Working Group did not have enough confidence in the proposed methods. The original AHP was heavily criticized in the scientific literature, and SMART was rarely mentioned during the workshops. It was unclear whether the ELECTRE method, the typical product of the French school in MCDA, would be suited to large-scale experiments. Thus, although SMART seemed to be the best candidate to support the choice of working languages, several methodological issues should have been settled before the decision process could begin.

In retrospect, we have good reasons to believe that our proposal touched a delicate issue in the European Community. That became clear in the summer of 1992. In June, the euphoria about the European unification came to an end. The Danish population rejected the Treaty of Maastricht with a tiny majority of 50.7%. Thereafter the French president Mitterand announced that he would submit the Treaty to the French population in order to accelerate the ratification process in the Community. In August 1992, shortly before the referendum, opinion polls came with the unexpected result that a majority of the French population would vote against Maastricht. The uncertain future of the Treaty triggered a valuta crisis in the Community, which prompted the authorities as well as the media in France and England to reveal their deeply hidden apprehension of a German domination. A tiny majority, 51% of the French population, eventually voted in favour of Maastricht. Nevertheless, confidence in the European unification was at a low level in the early nineties, whereas a surprising nationalism swept over Eastern Europe, despite a century of war and destruction in that part of the world. Nationalism may have been a crucial factor in the Working Group as well.

Nothing has really been solved in the years thereafter. There is a German Working Group now (DGOR Arbeitsgruppe "Entscheidungstheorie") with German and English as its working languages. Moreover, Roy and Vanderpooten (1996) recently declared the French school to be the European school of MCDA. In the comment on their paper the present author refuted this Napoleonic move. The French school is practically unknown outside the francophone area in Europe. A fortiori, however, why should we geographically distinguish several schools in MCDA (a French, an American, and now a European school) if they don't clearly correspond with the styles of decision making in the respective areas?

11.3 SCHOOLS ON ISLANDS

The formation of schools is a well-known phenomenon in the scientific world. Kuhn (1971) described it as a characteristic element of a pre-paradigm period. A paradigm, in his work, is the cognitive pattern of a scientific community, the collection of shared examples of how to conduct scientific research. The components of a paradigm are a shared symbolism for definitions and laws or relationships, a shared commitment to basic symbols and beliefs, and shared values of what constitutes a scientific result. Without a generally accepted paradigm there are only competing schools. Each author feels forced to build his work anew from its foundations. The pre-paradigm period is regularly marked by frequent and deep debates about legitimate methods, problems, and standards of solution, though these serve rather to define schools than to produce agreement. The dialogue of the resulting publications is often directed as much to the members of the school as it is to nature. This produces a pool of facts, a morass, and there is a constant reiteration of fundamentals.

This is exactly what happens in MCDA. We still don't have a shared view on how human preferences and values should be modelled, for instance, and we don't have a generally accepted idea of how decision processes in groups proceed via negotiations and power games. We only have an incoherent conglomerate of methods, techniques, and approaches, mostly for a single decision maker only. The schools of thought in MCDA operate in isolation, on separate islands, and there is only one ferry, the Journal of MCDA, which could possibly take care of the communication. It is not so easy for the editors, however, to maintain this role for the Journal. The referees may be rather intolerant whenever they have to judge the manuscripts from competing schools.

"How to handle manuscripts which discuss the basic assumptions of the schools in MCDA? Referees from a sympathetic school seem to be either very uncritical or, surprisingly, remarkably critical of technical detail, whereas referees from an unsympathetic school seem to argue against the methodology rather than to assess the quality of the paper". Simon French, General Editor of the Journal of MCDA, submitted this question to MCDA specialists and policy analysts who all came together (October 5, 1993) in the Group Decision Room (GDR) of the Faculty of Systems Engineering and Policy Analysis, Delft University of Technology, in order to experiment with advanced tools for group decision support.

"There are several possible policies", Simon French argued. *"Continue as present, with one referee from a sympathetic school and one from a different school. Alternatively, assign two referees from sympathetic schools only, or assign two referees from different schools; or offer the referees a page or two of discussion to be published along with any reply from the authors. In the last few months I considered these and other policies and I would be glad with your comments. Eventually, I hope to find the most workable policy. I do not immediately see the ideal mode of operation".*

The experiment started with an unstructured brainstorming session (for details, see Bots et al., 1994). The GroupSystemV of the GDR, which we also used for other experiments in later years (see Section 4.4 and Section 7.8), enabled the participants to input, at their own rate, their individual comments and suggestions from which the facilitator could easily extract a number of objectives for an editorial policy. After the lunch break the participants were asked to rank 16 objectives in a descending order of priority. Combination of the individually assigned rank-order positions produced a list of criteria in a rough group rank order. After a lively discussion about the mutual dependence of some objectives, the group finally accepted the following criteria to judge the alternative policies:

1. Enhancement of the scientific debate.

2. Enhancement of the quality of the papers.

3. Raising the number of readers.

4. Feasibility of the policy.

Initially, 21 alternative editorial policies were generated in the same way as the objectives. The individual rankings were combined into a group rank order, and after some discussion the group accepted the following list of policies for the final evaluation:

1. Choose referees from different, non-sympathetic schools.

2. Choose one referee from a sympathetic and one from a non-sympathetic school.

3. Use referee committees that judge several methodological papers at once.

4. Make special issues for the respective schools.

5. Publish the comments of the referees with the paper.

6. Let specific groups of specialists referee special methodological issues.

7. Focus on application domains rather than MCDA schools.

8. Give special instructions to non-sympathetic referees.

9. Narrow the scope of the journal.

The final evaluation, carried out via Leo Rog's implementation of SMART (Chapter 2), produced robust results. Alternative 5 (publication of the referees' comments) appeared to be high-ranking or dominant under all criteria so that it was the preferred alternative under large variations of the criterion weights. Alternative 2 (one sympathetic and one non-sympathetic referee) and Alternative 4 (special issues for the respective schools) ranked second and third. This result emerged from the session in the GDR, despite the time constraint, and Alternative 5 has indeed been implemented. The Journal of MCDA regularly publishes heated discussions between authors and referees.

The evaluation of the experiment itself, at the end of the day, triggered many critical comments. Although the participants generally agreed with the conclusions, they felt that there were several shortcomings in the formulation of the alternatives and the criteria, and also in the methodology. It is our experience that such an experiment should have been carried out in two rounds. The first round makes the participants familiar with the tools for decision support. Thereafter, some group members should sit down to redefine the alternatives and the criteria for the second round. In the next group meeting it is then much easier to complete the decision process in a satisfactory manner. It frequently happens, however, that the conclusions of two rounds almost coincide.

Let us, for instance, reconsider the alternatives. Several participants remarked that they were not mutually exclusive. One should consider strategies, each being a combination of tactics (the policies of the first round). Thus, if there would not have been a time constraint, the group could have considered the following higher-level alternatives.

1. An aggressive strategy, addressing the problem of non-communication and conflict between schools. Choose one or two referees from non-sympathetic schools, publish the comments of the referees, and make special issues of methodological papers where the controversies are openly discussed.

2. A strategy of containment, leaving it to later generations of scientists to discover the valuable ideas in each of the schools and to reject the inadequate assumptions and approaches. Choose referees from sympathetic schools only, and make special issues for the respective schools.

3. An evasive strategy, which avoids the problem of non-communication and conflict between schools. Thus, focus on applications, narrow the scope of the Journal, and choose referees from sympathetic schools only.

This might reopen the discussions, but now on a broader front, because one does not consider the individual tactics of a strategy, but the strategies themselves under the same or under redefined criteria.

As far as ease of use and conceptual simplicity are concerned, SMART seemed to be the best MCDA method for electronic group decision support. Well-known methods such as the AHP, MAUT, and ELECTRE are possibly too complicated and too time-consuming for GDR sessions, where the participants are usually under a heavy pressure.

REFERENCES TO CHAPTER 11

1. Bots, P., Kok, M., Lootsma, F.A., and Rog, L., "Schools on Islands, a Journal as a Ferry". *Journal of Multi-Criteria Decision Analysis* 3, 123 – 129, 1994.

2. French, S., *"Decision Theory, an Introduction to the Mathematics of Rationality"*. Ellis Horwood, Chichester, UK, 1988.

3. French, S., Simpson, L., Atherton, E., Belton, V., Dawes, R., Edwards, W., Hämäläinen, R., Larichev, O., Lootsma, F.A., Pearman, A., and Vlek, C., "Problem Formulation for Multi-Criteria Decision Analysis: Report of a Workshop". *Journal of Multi-Criteria Decision Analysis* 7, 242 – 262, 1998.

4. Kuhn, T.S., *"The Structure of Scientific Revolutions"*. Second Edition, University of Chicago Press, Chicago, Ill., 1971.

5. Lootsma, F.A., "The French and the American School in Multi-Criteria Decision Analysis". *RAIRO/Recherche Opérationnelle* 24, 263 – 285, 1990. A short version appeared in A. Goicoechea, L. Duckstein, and S. Zionts (eds.), *"Multiple Criteria Decision Analysis"*. Springer, Berlin, 1992, pp. 253 – 268.

6. Lootsma, F.A., "The Decision Maker and the Analyst in MCDA". *Journal of Multi-Criteria Decision Analysis* 5, 167 – 168, 1996.

7. Lootsma, F.A., Rog, L., and Kok, M., "Choice of Working Languages in a European Working Group for Multicriteria Decision Aid". *Journal of Information Science and Technology* 2, 170 – 176, 1993.

8. Mintzberg, H., *"Power in and around Organizations"*. Prentice-Hall, Englewood Cliffs, NJ, 1983.

9. Moore, G.A., *"Crossing the Chasm, Marketing and Selling Technology Products to Mainstream Customers"*. Harper Business, 1991.

10. Roy, B., *"Méthodologie Multicritère d'Aide à la Décision"*. Economica, Collection Gestion, Paris, 1985.

11. Roy, B., and Vanderpooten, D., "The European School of MCDA: Emergence, Basic Features, and Current Works". *Journal of Multi-Criteria Decision Analysis* 5, 22 – 37, 1996. In the same volume there is a comment by F.A. Lootsma (37 – 38) and a response by the authors (165 – 166).

12. Schärlig, A., *"Décider sur Plusieurs Critères"*. Presses Polytechniques Romandes, Lausannes, Switzerland, 1985.

13. Vincke, P., *"L'Aide Multicritère à la Décision"*. Editions de l'Université de Bruxelles, Brussels, Belgium, 1989.

14. Winterfeldt, D. von, and Edwards, W., *"Decision Analysis and Behavioral Research"*. Cambridge University Press, Cambridge, UK, 1986.

CHAPTER 12

PROSPECTS OF MCDA

The last chapter of a book usually contains the loose ends, the odd bits and pieces that have been left out of consideration in the previous chapters. So far, we have written so much about past experiences that it is time now to turn to the future of MCDA. What are its prospects, from the author's viewpoint, and what can be said about the desirable properties of the MCDA tools for decision support?

12.1 MCDA, A NEW TECHNOLOGY

MCDA is a new technology for decision making. The individual decision maker may resort to a pocket calculator to apply MCDA, but a group of decision makers, usually under a heavy pressure of time, needs a more sophisticated information system. The tools are commercially available indeed. Group decision rooms with networked PC's and a public screen for electronic brainstorming and voting have been installed in several establishments (LaPlante, 1993). Do these tools satisfy the needs of the decision makers? What kind of support would be welcome? So far, we heard only two answers: (a) savings in time and money, the objective of industry where the responsibility for time and money is paramount, and (b) improved consistency, a point of major concern in ministries because they have to justify their decisions *post hoc* in parliamentary sessions.

The history of technological innovations shows that many new products and new techniques were successful because they increased the capabilities of the users by an order of magnitude. Trains, steamboats, and motor cars accelerated travel and transport by a factor of 10 or more. In later years the products of the innovative aircraft industry had an impact of a similar order of magnitude. Telegraph and telephone increased the speed of communication even by several orders of magnitude. Radio and TV multiplied the audience of entertainers and politicians, and they opened the world for all classes of the population. In the nineteen fifties, sixties, and seventies, each new generation of computers was an order of magnitude faster than the previous one. Eventually, these techniques became indispensable. Life without them seems to be unthinkable now. True, not all innovations have such an attractive property, but the extraordinary increase in human productivity helps to carry the huge investment costs of the new technology. Nowadays, although the overall benefits are still difficult to quantify, information systems, automation, and robotics seem to increase the productivity of the labour force in such a manner that the high implementation costs are not prohibitive at all.

In the areas of Technological Forecasting (Lanford, 1972) and Technology Assessment (White, 1988) the introduction and penetration of new technologies as well as the substitution of conventional by new technologies has been thoroughly studied. These processes are lengthy. The substitution of horse-power and sail by steam took more than a century. Even now the take-off period from the initial scientific discovery to full-scale production and sales may take two or three decades. A well-known phenomenon over a long period of time is the emergence of a series of technologies to satisfy the demand for an improved capability (the candle, the whale-oil lamp, the kerosene lamp, and the incandescent lamp for lighting; the train, the motor car, and the aircraft for transportation). Any of these technologies seems to evolve along an S-shaped curve of capability improvement. The tangent to the peaks of the successive S-shaped performance curves exhibits an exponential improvement of the capability over a long period of technological progress.

Unfortunately, there is no indication that MCDA increases the capability or the productivity of the decision makers by an order of magnitude. Even after three decades of research, MCDA is still in the laboratory phase, despite the fact that some consulting firms now employ it in their commercial practice. The MCDA literature is full of isolated applications, but these case studies do not tell us whether MCDA will be a permanent, indispensable tool of management in the user's environment after the first successful experiment. Let us give a simple example in order to illustrate what we mean. A secretary who switches over from a typewriter to a text processor will probably tell, after a few weeks, that he/she would not like to return to the old machine. That is technological progress. It provides new tools which rapidly become indispensable. However, will MCDA become indispensable for decision making?

Direct savings of time and money in the decision process or an increased productivity of the decision makers (more decisions per day) have not been reported, at least to our knowledge. If they occur, they may be marginal. Even the mathematical and computational tools of Operations Research appeared to improve the efficiency of industrial operations by a few percent only. That may be a large amount of money, but in many cases the Operations Research tools complicated the task of management so much they were eventually abandoned (Lootsma, 1994). It occasionally happens that MCDA is felt to have tangible benefits because the selected alternative leads to unexpected savings. Similarly, an improved consistency in a series of related decisions may have persuasive financial advantages because it speeds up the decision processes in particular sectors of public administration. Our Ministry of Finance found MCDA important enough to promote a special software package for MCDA in the public domain. In general, however, measurable benefits of MCDA cannot immediately be promised to the users. The future of MCDA depends on its power to improve the quality of the decision and/or the decision process.

MCDA cannot show its benefits without the meta-power of higher authorities who definitely want to restructure the decision processes in the areas under their jurisdiction. Otherwise, any method for MCDA, whether it is supported by an advanced information technology or not, remains a tool for hobbyists only. Industrial companies, for instance, would (almost) never publish an annual report if they were not under the legal obligation to do so, and construction firms would (almost) never show a risk analysis of a hazardous installation if public authorities did not explicitly demand them to deliver it. Similarly, decision-making bodies will (almost) never show the criteria and the criterion weights employed in their decisions unless they are urged to do so. Thus, external forces are decisive factors when one tries to streamline the decision processes in industrial management and public administration via MCDA and other methods for decision analysis. I have seen a similar course of events in a large multinational company. Automation became successful there, as soon as the whole process of systems analysis, systems design, and implementation was tightly controlled by line management. The benefits of automation were difficult to quantify. There was a widespread feeling in the company, however, that automation would improve, not only the quality of the products and the productivity of the labour force, but also the job satisfaction of the workers. The decision to automate was not exclusively based upon rational arguments, but also on vision, expectations, and confidence. MCDA is clearly waiting for a similar belief among the authorities behind the possible users.

In recent years the future of MCDA became more promising indeed (the author, 1999). It increasingly happens that the decision makers in public administration have to account for their judgement. In many countries there are certain laws forcing the decision makers to show their judgemental data to the individuals who are affected by their decisions. The general public, on the other hand, becomes increasingly aware of these legal obligations and the resulting opportunities for effective control. Wenneras and Wold (1997), for instance, successfully appealed to the Freedom of the Press Act in Sweden in order to get access to the peer-review scores of the Swedish Medical Research Council. Thereafter, they could easily show that female scientists in the medical sector were disadvantaged in the competition for research grants. The significance of the actions of Wenneras and Wold cannot easily be overestimated. Backed up by legislation which forces the decision makers to reveal the scores underlying their decisions, MCDA may be a sharp weapon in the struggle for transparent and well-structured decision processes. It may be an effective tool to keep an eye on the allocation of research funds by National Science Foundations. In our country large infra-structural projects are subject to a law-enforced Environmental

Assessment Procedure. This is in fact a multi-criteria evaluation of the alternative solutions for certain infra-structural bottlenecks (alternative motorway and railway trajectories, alternative airport extensions, river-dike improvements, etc.). The original as well as the judgemental data are publicly available. In summary, it is highly plausible that MCDA will regularly be used at all levels of (public) decision making.

12.2 THE REMBRANDT SYSTEM

The REMBRANDT system which we used in our experiments is not commercially available. Designed by Leo Rog (Delft University of Technology) and heavily tested by Olson et al. (1995) in a comparative study of several tools for MCDA, see also Olson (1996), it has been used to try out methodological ideas. Moreover, it was always freely available on request. In recent years, the Ministry of Finance invited us to consider a possible amalgamation of the REMBRANDT system and the DEFINITE program (BOSDA in Dutch) designed by Janssen and Herwijnen (1994) for the environmental assessment of alternative infrastructural (re)constructions via a variety of MCDA methods. The discussions, still without firm conclusions, prompted us to formulate extensively the desirable features of a system for decision support via MCDA.

Let us first summarize what triggered the discussions. The DEFINITE program, promoted by the Ministry of Finance since the early nineties, has been used by various public authorities and particularly by the Environmental Assessment Committee which has the legal task to review alternative project proposals for the (re)construction of railways and motorways, for the improvement of river dikes, and the like. The critical comments of the decision makers revealed, not the shortcomings of the DEFINITE program itself, but rather the unresolved issues in the field of MCDA.

1. There are five MCDA methods in the DEFINITE program (cardinal and ordinal ones), with five so-called standardization methods to evaluate the alternatives under the respective criteria, and with five methods to obtain criterion weights. The documentation does not tell the decision makers how to find a proper combination of methods, and honestly, the MCDA literature in general does not really help the decision makers either. A delicate point is that the criterion weights must be adjusted whenever the decision makers turn from one of the standardization methods to another. The documentation does not tell them how to carry out the adjustment. In general, the criterion weights depend on the performance of the alternatives, so that DEFINITE cannot easily be used for distributed decision making.

2. Each standardization method in the DEFINITE program is in fact an assessment of the alternatives under a given criterion, without a wider context. The original performance data, usually in physical or monetary units, are employed to assign a value between zero and one to each alternative. The worst and the best alternative in

the actual decision problem usually play an important role. Hence, the final rank order of the alternatives may be affected by the introduction of a new alternative which, under at least one of the criteria, is worse than the worst one or better than the best one so far. The standardization procedures are objective, however, in the sense that they only use the original performance data. Asking the technical experts to model the context explicitly would be an appeal to their subjective judgement, and this is not always welcome in public decision processes.

3. Group decisions are not supported in DEFINITE. The program has been designed for a single decision maker, or for an analyst who distillates a group opinion out of the group's judgemental statements. This implies that the group is considered to be a new, single decision maker (compromises throughout the analysis, see Chapter 6), not a collection of individual decision makers (compromises at the end of the analysis).

4. A hierarchy of criteria, subcriteria, etc., became very popular in recent years, particularly by the EXPERT CHOICE program incorporating the original AHP. In DEFINITE, however, there are only two evaluation levels: one for the criteria and one for the alternatives.

The features of the REMBRANDT system are mainly due to our experiences in The Netherlands and in Brussels. From the late seventies we have been confronted with group decision making and with the problem of how to find an acceptable compromise, despite the inhomogeneous power distribution in the group. Moreover, we were in close contact with decision-making bodies and advisory councils involved in a complicated process of distributed decision making.

1. REMBRANDT works with two cardinal MCDA methods only: the Multiplicative AHP with pairwise comparisons, as well as SMART with direct rating which is also known as the assignment of grades (the Additive AHP could easily be implemented as well). There is a logarithmic relationship between the two methods: differences of grades in SMART represent logarithms of preference ratios in the Multiplicative AHP. In fact, we employed logarithmic coding here, a familiar mode of operation in acoustics. This also explains what the acronym REMBRANDT means: Ratio Estimation in Magnitudes or deci-Bells to Rank Alternatives which are Non-DominaTed. The two methods, the Multiplicative AHP and SMART, are thoroughly mixed. Under each criterion the decision maker has the choice: he/she may employ direct rating or pairwise comparisons. The concept of the relative importance of the criteria is based upon the scale which is used in direct rating and in the method of pairwise comparisons, not on the alternatives which happen to be present in the actual decision problem (see Chapter 2 and Chapter 3). REMBRANDT may therefore be used in distributed decision making, when there is an organizational distance between the evaluation of the criteria and the assessment of the alternatives. The criterion weights may even be transferred from previous, similar decision problems to the actual one, so that there is an enhanced consistency in a coherent series of mutually related decisions.

2. In REMBRANDT the assessment of the alternatives under a given criterion is in principle carried out within a predetermined context, which may or may not explicitly be formulated by the decision maker. In direct rating, we recommend the decision maker to sketch explicitly the real or imaginary alternatives corresponding to the endpoints of the scale (the excellent alternatives and those with a poor but compensatable performance). In the application of pairwise comparison methods the context will intuitively be present, but usually we recommend the decision makers to describe it explicitly as well.

3. Group decisions are extensively supported by REMBRANDT. The system accepts the judgemental statements of each group member individually, and it shows each member's terminal scores of the alternatives. The power distribution is modelled via weights explicitly assigned to the decision makers. The search for a compromise solution is supported by the calculation of weighted group terminal scores of the alternatives, under the assumption that the group acts as a new, single decision maker (compromises throughout the analysis). A revised version of REMBRANDT could also support the search for a compromise in situations where the group is considered to be a collection of individuals (compromises at the end of the analysis).

4. A hierarchy of criteria, subcriteria, etc, cannot be analyzed in REMBRANDT. A hierarchical structure with more than two evaluation levels should be thoroughly studied before it is launched in a practical environment. In Chapter 2 and Chapter 3 we only formalized the concept of the relative importance of the criteria. In a hierarchy with more than two evaluation levels one runs up against the concept of the relative importance of the subcriteria, the sub-subcriteria, etc., concepts which are still undefined. Moreover, the structure of the hierarchy is not always uniquely determined by the problem formulation, a phenomenon which strongly affects the weights in the analysis, see Section 3.7. The original AHP disregards these questions and constructs multi-level hierarchies as audaciously as it carries out the analysis thereafter.

In their present form, both DEFINITE and REMBRANDT are simple programs which do not fully use the graphical facilities of Windows. They have not been installed in Group Decision Rooms either. In our experiments with groups of decision makers, Leo Rog added some *ad hoc* implementations of SMART to the GroupsystemV software of our Group Decision Room. The benefits and the pitfalls of MCDA in real-life electronic brainstorming and voting are still on the research agenda.

12.3 THE UNBEARABLE LIGHTNESS OF IMPRECISION

The success of MCDA depends critically on the position of the decision makers or actors within the framework of the decision process. It is hazardous to apply MCDA when the decision makers do not have the power or the authority to implement their decisions and when they are not responsible for the consequences. This is typically the situation encountered in advisory councils. They have a certain amount of influence, and unanimity consolidates their position, but on many occasions they are only requested to consider the decision problem from a particular viewpoint and with a particular expertise. They do not always see the problem in its totality, so that their advice may easily be rejected or ignored. The position of an advisory group is not always precisely defined. In many organizations the advice of a nomination committee, for instance, is practically binding. Nobody is prepared to specify the conditions under which the advice may be rejected. Both, the committee and those who established the committee, have to find an acceptable mode of operation in the twilight zone, and they usually do.

Many decisions are so complicated that they must be prepared in a network of technical and political committees with an advisory position (distributed decision making). Infrastructural decisions such as the choice of a high-speed railway trajectory are eventually made in the cabinet, after many years of preparation at municipal, provincial, and national levels. Sometimes, the decision makers or actors in the advisory committees do not properly see their role. A major point of concern is the choice of the criteria. Where and when, and by whom will the choice be made? Does it make sense to discuss the relative importance of the criteria before the alternatives are known? What are the relevant scenarios of future developments? How to present or to summarize the assessment of the alternatives before the real decision makers are known?

The success of MCDA also depends on the hidden agenda of the decision makers. If they do not want to achieve consensus, or if they do not even want to take a decision at all, MCDA can rapidly identify the divergent opinions in the group, it can easily show that the relative importance of the criteria varies from subgroup to subgroup, but it cannot impose a compromise solution. A decision sustained by a majority subgroup or by an *ad hoc* coalition only emerges under the pressure of external forces.

Sometimes, one has to warn the decision makers against the precision of numbers in MCDA. The numbers usually represent judgemental categories only. The grades in SMART and the scale values in the AHP are typical examples. The grade 10 stands for "excellent" in various gradations, the grade 8 for "good" in various gradations, etc. Jointly, the grades cover the range of acceptable performance. In a certain sense, they are like colour terms which represent vaguely defined subsets of the three-dimensional space of hue, brightness, and saturation, and which jointly cover the colour sphere. Certain arithmetic operations in MCDA, like the calculation of means of grades, produce helpful intermediate results. In various stages of the decision process, however, the decision makers have to relax in order to ask themselves whether they agree with the calculated results at a deeper level of understanding.

The decision makers are usually aware of the imprecision of words. Many policy documents, written in an administrative or managerial jargon, are full of terms with a quantitative connotation, like "somewhat more", "many",.... The authors avoid any further specification. We therefore usually suggest the decision makers to specify the context of the decision problem as far as possible, and to search for a plausible interpretation of these quantifiers within the actual context.

Nevertheless, imprecision is unavoidable in decision making. One cannot and one does not have to refine categorical human judgement and the associated scales. Decision makers are not in the position of car manufacturers who have to refine the vague colour terms into precisely defined colour codes. Imprecision may even be welcome. It lubricates human communication and cooperation. The ratios assigned to pairs of research output items (Section 7.8), for instance, are prototypical numbers or numerical labels matching the judgement of the decision makers. The contributions of the output items to the advancement of pure and/or applied science, to innovative problem solving, ..., cannot precisely be defined, certainly not because the faculties in a university have strongly divergent views and experiences. In actual decision making, however, one only needs a rough indication. It will be easier for the faculties to agree on such a vague indication than on a precisely established numerical value.

Eventually, the moment of truth arrives for each member of a decision-making body or an advisory council, when they individually have to make up their mind for the final decision. It may be almost midnight. Many alternatives have already been dropped, many criteria have been set aside as irrelevant. Then it happens that the grades and scores reveal their message in a light and easy manner. What emerges is not only the preferred alternative but also the required commitment. The decision maker feels that the cells in his/her body seek their proper orientation until they are lined up to carry out the decision.

REFERENCES TO CHAPTER 12

1. Janssen, R., and Herwijnen, M., *"Definite, a System to Support Decisions on a FINITE Set of Alternatives"*. Kluwer, Dordrecht/Boston/London, 1994.

2. Lanford, H.W., *"Technological Forecasting Methodologies, a Synthesis"*. American Management Association, 1972.

3. LaPlante, A., "Nineties Style Brainstorming". *Technology Supplement of Forbes Magazine,* October 25, 44 – 61, 1993.

4. Lootsma, F.A., "Alternative Optimization Strategies for Large-Scale Production Allocation Problems". *European Journal of Operational Research* 75, 13 – 40, 1994.

5. Lootsma, F.A., "MCDA, a New Technology". *Journal of Multi-Criteria Decision Analysis* 5, 245 – 246, 1996.

6. Lootsma, F.A., "The Expected Future of MCDA". To appear in the *Journal of Multi-Criteria Decision Analysis,* 1999.

7. Olson, D., *"Decision Aids for Selection Problems"*. Springer Series in Operations Research, New York, 1996.

8. Olson, D., Fliedner, G., and Currie, K., "Comparison of the REMBRANDT System with the AHP". *European Journal of Operational Research* 82, 522 – 539, 1995.

9. Wenneras, W., and Wold, A., "Nepotism and Sexism in Peer-Review". *Nature* 387, 341 – 343, 1997.

10. White, B.L., *"The Technology Assessment Process"*. Quorum, New York, 1988.

[illegible] F[illegible] the E[illegible] of MC[illegible] [illegible]

Olson, D. [illegible] [illegible]

Olson, [illegible] O. [illegible] Comparison of the [illegible] System with [illegible] 1995.

[illegible] and [illegible], A. [illegible]

[illegible] E. [illegible]

SUBJECT INDEX

ABOUT THE AUTHOR

Freerk Auke Lootsma, born in The Netherlands (1936), Gymnasium B (sciences, 1953), Gymnasium A (humanities, 1954), studied mathematics and theoretical physics at the State University of Utrecht (1954 – 1961). He presented a Ph.D. dissertation on non-linear optimization at the Eindhoven University of Technology (1970). Enrolled in the Royal Netherlands Navy he was introduced into Operations Research at the National Defence Research Laboratory in The Hague (1961 – 1962). As a scientific staff member of Philips Research Laboratories in Eindhoven (1963 – 1970) and in Redhill, Surrey, UK (1970 – 1971) he was mainly involved in business applications of Operations Research such as production allocation and scheduling, the trim-loss problem, and traffic control. In Philips' Information Systems and Automation Department and in the Management Training Centre he studied the automation policy of the Philips Concern, in cooperation with members of Harvard Business School (1972 – 1974). Since 1974 he is a senior professor of the Delft University of Technology, first in the Faculty of Mathematics and Informatics, since 1997 in the Faculty of Information Technology and Systems. His main interests are in Operations Research and Decision Analysis, with applications in energy planning at the national and the European level (projects in The Hague and Brussels) and in public health. He directed or co-directed NATO Advanced Research and Study Institutes in Urbino, Italy (1977), Cambridge, UK (1981), and Val d'Isère, France (1987). In addition he held several administrative positions: Treasurer of the Mathematical Programming Society (1974 – 1980), Chairman of his Department (1980 – 1982, 1988 – 1990, 1994 – 1996), Vice-Chancellor of his University (1984 – 1988), and Board Member of the Computing Centre of the Dutch Reformed Church (1978 – 1985). He spent the academic year 1993 – 1994 on sabbatical leave at the Department of Mechanical Engineering and Applied Mechanics, University of Michigan, Ann Arbor, USA.

ABOUT THE AUTHOR

Applied Optimization

1. D.-Z. Du and D.F. Hsu (eds.): *Combinatorial Network Theory.* 1996 ISBN 0-7923-3777-8
2. M.J. Panik: *Linear Programming: Mathematics, Theory and Algorithms.* 1996 ISBN 0-7923-3782-4
3. R.B. Kearfott and V. Kreinovich (eds.): *Applications of Interval Computations.* 1996 ISBN 0-7923-3847-2
4. N. Hritonenko and Y. Yatsenko: *Modeling and Optimimization of the Lifetime of Technology.* 1996 ISBN 0-7923-4014-0
5. T. Terlaky (ed.): *Interior Point Methods of Mathematical Programming.* 1996 ISBN 0-7923-4201-1
6. B. Jansen: *Interior Point Techniques in Optimization.* Complementarity, Sensitivity and Algorithms. 1997 ISBN 0-7923-4430-8
7. A. Migdalas, P.M. Pardalos and S. Story (eds.): *Parallel Computing in Optimization.* 1997 ISBN 0-7923-4583-5
8. F.A. Lootsma: *Fuzzy Logic for Planning and Decision Making.* 1997 ISBN 0-7923-4681-5
9. J.A. dos Santos Gromicho: *Quasiconvex Optimization and Location Theory.* 1998 ISBN 0-7923-4694-7
10. V. Kreinovich, A. Lakeyev, J. Rohn and P. Kahl: *Computational Complexity and Feasibility of Data Processing and Interval Computations.* 1998 ISBN 0-7923-4865-6
11. J. Gil-Aluja: *The Interactive Management of Human Resources in Uncertainty.* 1998 ISBN 0-7923-4886-9
12. C. Zopounidis and A.I. Dimitras: *Multicriteria Decision Aid Methods for the Prediction of Business Failure.* 1998 ISBN 0-7923-4900-8
13. F. Giannessi, S. Komlósi and T. Rapcsák (eds.): *New Trends in Mathematical Programming.* Homage to Steven Vajda. 1998 ISBN 0-7923-5036-7
14. Ya-xiang Yuan (ed.): *Advances in Nonlinear Programming.* Proceedings of the '96 International Conference on Nonlinear Programming. 1998 ISBN 0-7923-5053-7
15. W.W. Hager and P.M. Pardalos: *Optimal Control.* Theory, Algorithms, and Applications. 1998 ISBN 0-7923-5067-7
16. Gang Yu (ed.): *Industrial Applications of Combinatorial Optimization.* 1998 ISBN 0-7923-5073-1
17. D. Braha and O. Maimon (eds.): *A Mathematical Theory of Design: Foundations, Algorithms and Applications.* 1998 ISBN 0-7923-5079-0

Applied Optimization

18. O. Maimon, E. Khmelnitsky and K. Kogan: *Optimal Flow Control in Manufacturing.* Production Planning and Scheduling. 1998 ISBN 0-7923-5106-1
19. C. Zopounidis and P.M. Pardalos (eds.): *Managing in Uncertainty: Theory and Practice.* 1998 ISBN 0-7923-5110-X
20. A.S. Belenky: *Operations Research in Transportation Systems:* Ideas and Schemes of Optimization Methods for Strategic Planning and Operations Management. 1998 ISBN 0-7923-5157-6
21. J. Gil-Aluja: *Investment in Uncertainty.* 1999 ISBN 0-7923-5296-3
22. M. Fukushima and L. Qi (eds.): *Reformulation: Nonsmooth, Piecewise Smooth, Semismooth and Smooting Methods.* 1999 ISBN 0-7923-5320-X
23. M. Patriksson: *Nonlinear Programming and Variational Inequality Problems.* A Unified Approach. 1999 ISBN 0-7923-5455-9
24. R. De Leone, A. Murli, P.M. Pardalos and G. Toraldo (eds.): *High Performance Algorithms and Software in Nonlinear Optimization.* 1999 ISBN 0-7923-5483-4
25. A. Schöbel: *Locating Lines and Hyperplanes: Theory and Algorithms.* 1999 ISBN 0-7923-5559-8
26. R.B. Statnikov: *Multicriteria Design.* 1999 ISBN 0-7923-5560-1

KLUWER ACADEMIC PUBLISHERS – DORDRECHT / BOSTON / LONDON